Hormone Research

International Journal of Experimental and Clinical Endocrinology
Founded 1970 as 'Hormones' by M. Marois

Official Organ of the European Society for Paediatric Endocrinology

General Information

Publication data: 'Hormone Research' is published 12 times annually. Volumes 33–34 with 6 issues each appear in 1990

Subscription rates: Subscriptions run for a full calendar year. Prices are given per volume, surface postage included.
Personal subscription: SFr. 245.–, US$ 149.10. (Must be in the name of, billed to, and paid by an individual. Order must be marked 'personal subscription'.)
Institutional subscription: SFr. 350.–, US$ 213.00. (Regular rate.)
Members of the European Society of Paediatric Endocrinology: SFr. 67.20, US$ 42.60.
Lawson Wilkins Society; Schweizerische Gesellschaft für Endokrinologie: SFr. 210.–, US$ 127.80.
Airmail postage: SFr. 35.–, US$ 25.00. (Extra per volume.)
Microform subscription: SFr. 175.–, US$ 106.50. (Available exclusively to subscribers of the paper edition.)

Single issues and back volumes: Information on availability and prices can be obtained through the Publisher.

Subscription orders: Orders can be placed at agencies, bookstores, directly with the Publisher S. Karger AG, P.O. Box, CH – 4009 Basel (Switzerland) or with any of the following distributors:
Deutschland, West-Berlin, Österreich, Ost-Europa: S. Karger GmbH, Postfach 1724 D – 8034 Germering/München
France: Librairie Luginbühl 36, bd de Latour-Maubourg, F – 75007 Paris
Great Britain, Ireland: S. Karger Publishers, 1 Torberry Drive Petersfield, Hampshire GU 31 4HW
USA: S. Karger Publishers, Inc. 26 West Avon Road, P.O. Box 529 Farmington, CT 06085
India, Bangladesh, Sri Lanka: S. Karger India Subscription Agency B – 5/132 Safdarjung Enclave New Delhi-110029
Japan: Katakura Libri, Inc. 36-9 Hongo 3-chome, Bunkyo-ku, Tokyo 113

Change of address: Both old and new address should be stated and sent to the subscription source.

Bibliographic indices: This journal is regularly listed in bibliographic services, including *Current Contents®*

Book Reviews: Books are accepted for review by special agreement.

Advertising: Correspondence and rate requests should be addressed to the Publisher.

KARGER

Horm Res
Printed in Switzerland

S. Karger AG,
Postfach, CH – 4009 Basel (Schweiz)

Erscheint 12mal jährlich

'Hormone Research' aims to provide rapid publication of clinical investigations and case reports, as well as experimental and methodological work related to endocrinology. Correspondence is welcome, but publication of letters is at the discretion of the editors.

Submission: Only original papers written in English are considered and should be sent to:
Prof. J. Girard
Universitäts-Kinderklinik Basel
Römergasse 8
CH-4005 Basel (Switzerland)
Manuscripts should be submitted in triplicate (with three sets of illustrations of which one is an original), typewritten double-spaced on one side of the paper, with a wide margin.

Conditions: All manuscripts are subject to editorial review. Manuscripts are received with the explicit understanding that they are not under simultaneous consideration by any other publication. Submission of an article for publication implies the transfer of the copyright from the author to the publisher upon acceptance. Accepted papers become the permanent property of 'Hormone Research' and may not be reproduced by any means, in whole or in part, without the written consent of the publisher. It is the author's responsibility to obtain permission to reproduce illustrations, tables, etc. from other publications.

Arrangement:

Title page: The first page of each paper should indicate the title (main title underlined), the authors' names, and the institute where the work was conducted. A short title for use as running head is also required.

Key words: For indexing purposes, a list of 3–10 key words in English is essential.

Abstract: Each paper needs an abstract of up to 10 lines.

Small type: Paragraphs which can or must be set in smaller type (case histories, test methods, etc.) should be indicated with a 'p' (petit) in the margin on the left-hand side.

Footnotes: Avoid footnotes. When essential, they are numbered consecutively and typed at the foot of the appropriate page.

Tables and illustrations: Tables and illustrations (both numbered in Arabic numerals) should be prepared on separate sheets. Tables require a heading and figures a legend, also prepared on a separate sheet. For the reproduction of illustrations, only good drawings and original photographs can be accepted; negatives or photocopies cannot be used. When possible, group several illustrations on one block for reproduction (max. size 176 × 224 mm) or provide crop marks. On the back of each illustration, indicate its number, the author's name, and 'top'. Color illustrations are reproduced at the author's expense.

References: In the text identify references by Arabic numerals [in square brackets]. Material submitted for publication but not yet accepted should be noted as 'unpublished data' and not be included in the reference list. The list of references should include only those publications which are cited in the text. Do not alphabetize; number references in the order in which they are first mentioned in the text. The surnames of the authors followed by initials should be given. There should be no punctuation other than a comma to separate the authors. Cite all authors, 'et al' is not sufficient. Abbreviate journal names according to the Index Medicus system. (Also see International Committee of Medical Journal Editors: Uniform requirements for manuscripts submitted to biomedical journals. Br Med J 1988;296:401–405.)

Examples

(a) Papers published in periodicals: Kauffman HF, van der Heide S, Beaumont F, Blok H, de Vries K: Class-specific antibody determination against *Aspergillus fumigatus* by means of the enzyme-linked immunosorbent assay. III. Comparative study: IgG, IgA, IgM ELISA titers, precipitating antibodies and IgE binding after fractionation of the antigen. Int Arch Allergy Appl Immunol 1986;80:300–306.

(b) Monographs: Matthews DE, Farewell VT: Using and Understanding Medical Statistics. Basel, Karger, 1985.

(c) Edited books: Hardy WD Jr, Essex M: FeLV-induced feline acquired immune deficiency syndrome: A model for human AIDS; in Klein E (ed): Acquired Immunodeficiency Syndrome. Prog Allergy. Basel, Karger, 1986, vol 37, pp 353–376.

Full address: The exact postal address complete with postal code of the senior author must be given. If correspondence is handled by someone else, indicate this accordingly.

Further details: Authors may consult our leaflet 'Rules for the Preparation of Manuscripts', available on request from the Publisher, S. Karger AG, P.O. Box, CH-4009 Basel (Switzerland).

Page charges: There is no page charge for papers of 4 or fewer printed pages (including tables, illustrations and references). Each additional complete or partial page is charged to the author at SFr. 415.– / US$ 275.00. The allotted size of a paper is equal to approx. 12 manuscript pages (including tables, illustrations and references).

Galley proofs: Unless indicated otherwise, galley proofs are sent to the first-named author and should be returned with the least possible delay. Alterations made in galley proofs, other than the correction of printer's errors, are charged to the author. No page proofs are supplied.

Reprints: Order forms and a price list are sent with the galley proofs. Orders submitted after the issue is printed are subject to considerably higher prices. Allow five weeks from date of publication for delivery of reprints.

KARGER

Pediatric and Adolescent Endocrinology (ISSN 0304–4254), **Vol. 19**
Series Editor: Z. Laron, Petah Tikva/Tel Aviv

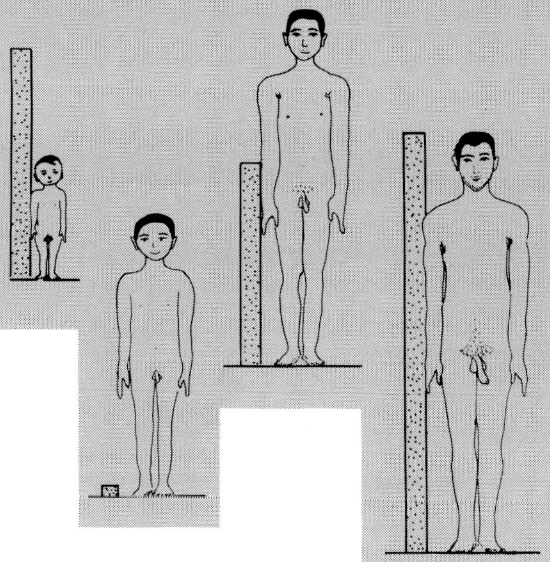

Androgens in Childhood

Editor:
M.G. Forest, Lyon

Biological, Physiological, Clinical and Therapeutic Aspects

The first in-depth guide to the subject

This volume brings together for the first time all recent knowledge concerning androgens in childhood. It features not only the basic biochemical background, the regulation of organs producing androgens, the mode of action and the ontogeny of androgen production, but also covers modern approaches to the pathology and management of androgen disorders in children and the use of androgen therapy in pediatric practice.
Authored by an international group of invited specialists in the field, this collection of papers represents a thorough and up-to-date handbook on the biology and physiology of androgens and new therapeutic approaches of immense value for pediatricians, endocrinologists and biochemists.

Fields of Interest

Biochemistry; Pediatrics; Andrology, Biology for Physicians, Endocrinology, Gynecology, Metabolism, Oncology, Pathology, Pharmacology, Physiology

Contents

Pediatric and Adolescent Endocrinology, Vol. 19
Androgens in Childhood
Biological, Physiological, Clinical and Therapeutic Aspects
Editor: Forest, M.G. (Lyon)
VIII + 282 p., 48 fig., 23 tab., hard cover, 1989
SFr. 224.– / DM 268.– / £ 93.40 / US$ 149.50
US$ / £ prices for USA / UK only (subject to change)
ISBN 3–8055–4850–8

KARGER

KI 89509

Pediatric and Adolescent Endocrinology (ISSN 0304–4254), **Vol. 18**
Series Editor: Z. Laron, Petah Tikva/Tel Aviv

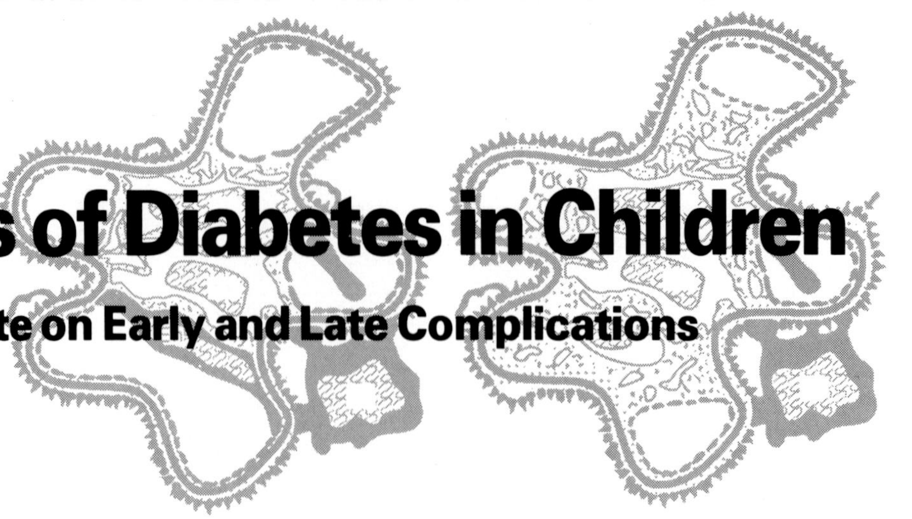

Prognosis of Diabetes in Children

An Update on Early and Late Complications

Editors:
Z. Laron, M. Karp,
Petah Tikva/Tel Aviv

The course of diabetes in childhood is accompanied by a variety of complications, some of which may appear relatively early, resulting in young adults with a marked handicap. The 43 contributions in this volume record the 7th International Beilinson Symposium, where a group of leading investigators exchanged the results of their ongoing inquiries on the pathogenesis of the complications of diabetes in children, the potential of recent medical advances for preventing these complications, and on the definition of early markers of their development. Further, evidence for correlations between diabetes control, genetic factors, environmental conditions and diabetes complications are reviewed. Taken together, these papers are a valuable reference for diabetologists, pediatricians, endocrinologists and internists.

A critical update of the multifaceted problems related to diabetic complications in children, with an emphasis on early detection and prevention

Contents

Introductory Lectures
Contributing authors: *Åkerblom, H.K.; Williamson, J.R., et al.;* The DCCT Research Group represented by Siebert, C.

Complications in the First Years of Diabetes
Contributing authors: *Fainmesser, P., et al.; de Beaufort, C.E., et al.; Hermansson, G., et al.; Jenkins, J.G., et al.; Tubiana-Rufi, N., et al.; Mukamel, M., et al.*

Neuropathy
Contributing authors: *Sima, A.A.F., et al.; Aagenaes, O.; Noel, P.; Käär, M.-L.*

Retinopathy
Contributing authors: *Standl, E.; Williamson, J.R.; Yassur, Y.; Dorchy, H., et al.; Weber, B.; Chantelau, E., et al.; Kernell, A., et al.*

Nephropathy
Contributing authors: *Borch-Johnsen, K.; Mathiesen, E.R., et al.; Wartha, R., et al.; Roe, T.F., et al.; Urakami, T., et al.; Chiarelli, F., et al.; Cantini, M., et al.*

Mortality
Contributing authors: *Tuomilehto, J.; Lounamaa, R. and Tuomilehto, J.*

Markers of Vascular Complications
Contributing authors: *Koepp, P. and Seeler, M.; Roth, J., et al.; Danne, T., et al; Monciotti, C., et al.*

Epidemiology
Contributing authors: *Kalter-Leibovici, O., et al.; Matsuura, N., et al.; Wang H.; Ko, K.W.; Davies, A.G., et al.*

Psychological Aspects
Contributing authors: *Galatzer, A., et al.; Ludvigsson, J. and Richt, B.; Kemmer, F.W.L., et al.*

Recent Advances
Contributing authors: *Bingley, P.J., et al.; Wolf, E.; Smith, C.P., et al.; Beyer, J., et al.; Zini, D., et al.; Hogan, M., et al.*

Fields of Interest

Biochemistry; Pediatrics; Metabolic Diseases; Angiology, Diabetes, Endocrinology, Metabolism, Nephrology, Neurology, Ophthalmology

Pediatric and Adolescent Endocrinology, Vol. 18
Prognosis of Diabetes in Children
An Update on Early and Late Complications
7th International Beilinson Symposium, Jerusalem, November 1987
Editors: Laron, Z.; Karp, M. (Petah Tikva/Tel Aviv)
XIV + 306 p., 68 fig., 86 tab., hard cover, 1989
SFr. 249.– / DM 298.– / £ 103.80 / US$ 166.00
US$ / £ prices for USA / UK only (subject to change)
ISBN 3–8055–4702–1

KARGER

KI 89508

Third Joint Meeting of the ESPE and LWPES

Third Joint Meeting of the European Society for Paediatric Endocrinology (ESPE) and the Lawson Wilkins Pediatric Endocrine Society (LWPES)

Jerusalem, October 29–November 3, 1989

Mode of Action of Hormones
hGH Therapy
Autoimmune Aspects of IDDM

Proceedings of the Plenary Lectures and Symposia

Guest Editor: Z. Laron, Tel Aviv, Israel

44 figures, 16 tables, 1990

KARGER

S. Karger · Medical and Scientific Publishers
Basel · München · Paris · London · New York · New Delhi · Bangkok · Singapore · Tokyo · Sydney

Drug Dosage

The authors and the publisher have exerted every effort to ensure that drug selection and dosage set forth in this text are in accord with current recommendations and practice at the time of publication. However, in view of ongoing research, changes in government regulations, and the constant flow of information relating to drug therapy and drug reactions, the reader is urged to check the package insert for each drug for any change in indications and dosage and for added warnings and precautions. This is particularly important when the recommended agent is a new and/or infrequently employed drug.

Printed in Switzerland by Thür AG Offsetdruck, Pratteln
ISBN 3-8055-5289-0

Contents

IV. Update on Growth Hormone Therapy

V. Update on Immunological Aspects of IDDM

Epilogue

Foreword

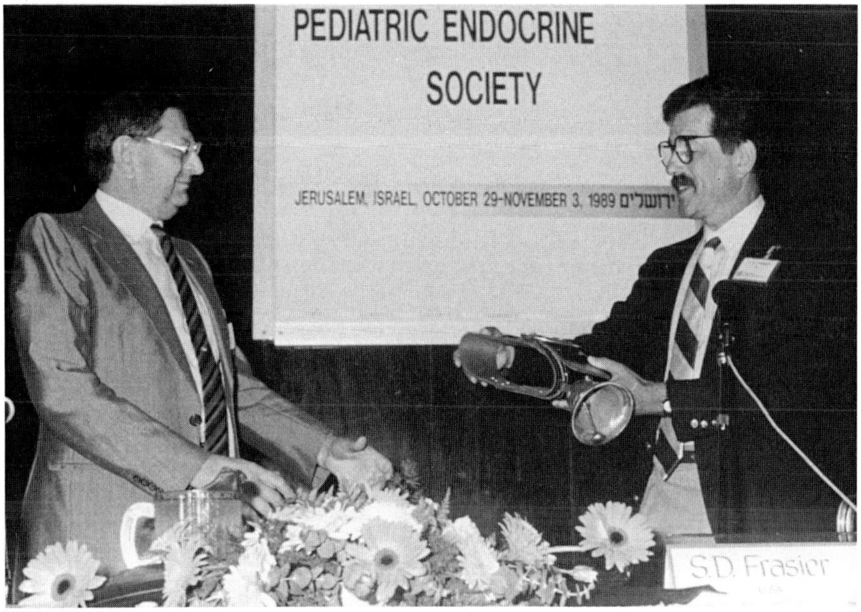

Dr. Douglas S. Frasier, President of the LWPES, hands over the Joint Meeting Cow Bell to Dr. Zvi Laron, President of the ESPE.

This volume of *Hormone Research* presents the Proceedings of the Third Joint Meeting of the European Society for Paediatric Endocrinology (ESPE) and the Lawson Wilkins Pediatric Endocrine Society (LWPES), held in Jerusalem between October 29 and November 3, 1989. These Joint Meetings are held every 4 years. The two previous ones were held in Geneva in 1981 and in Baltimore in 1985.

As President of the ESPE, I had the honor and pleasure of hosting this Third Joint Meeting. It became the largest encounter of pediatric endocrinologists held to date, with 776 participants from 39 countries. The scientific program lasted 3 full days and comprised 4 plenary lectures, 4 symposia with 16 speakers, 74 short communications and 218 poster presentations.

It was also the first official meeting of the representatives of the six now existing pediatric endocrine societies and the founding venue of the Coordination Office of these societies.

Many friendships were renewed and new ones formed during the social programs.

The proceedings of the volume have been rearranged into 5 sections:

I. Plenary lectures including the ESPE Prader Prize Lecture and the Lawson Wilkins Memorial Lecture.
II. Hormone Action at the Receptor and Postreceptor Level.
III. Mode of Action of Growth Hormone and IGF-I.
IV. Update on Growth Hormone Therapy.
V. Update on Immunological Aspects of IDDM.

It is hoped that these full reports and ensuing discussions as well as the 312 abstracts [Horm Res 1989; 31(suppl 1)] will serve as a reminder to those who have attended the Meeting and as a source of update information for those who were unable to join us.

I take this opportunity again to express our thanks to the many contributors to the high scientific level of the Third Joint Meeting, to the generous sponsors and to the Kenes Organization. I wish also to acknowledge the invaluable help of Mrs. Gila Waichman in the preparation of the meeting and in the editing of this volume.

Zvi Laron, MD
Irene and Nicholas Marsh
Professor of Pediatric Endocrinology and Diabetes
Sackler Faculty of Medicine
Tel Aviv University
Beilinson Medical Center, Israel

I. Plenary Lectures

Horm Res 1990;33:53–57

0301–0163/90/0334–0053$2.75/0

Chronobiology

W.J. Rietveld

Department of Physiology, University of Leiden, The Netherlands

Key Words. Chronobiology · Biological rhythmicity · Development

Abstract. The field of chronobiology, the study of the rhythms in plants and animals, was restricted to botanists for centuries. Only recently during the last decades could research be broadened to include animals and later even human beings. Rhythms have been documented and related to the alternation of day and night and to the succession of the seasons. Nowadays, chronobiology has developed into a multidisciplinary field in which scientists are involved in basic research as well as in applied topics. This paper gives an introduction to the field, especially dealing with the aspect of rhythm development, and the way in which the different 24-hour rhythms in children become apparent.

Introduction

Four or five billion years ago the first primitive form of life evolved from the constituents of the primeval broth. In interplay with the environment it has developed into the animal and vegetable kingdoms as we know them today.

The fact that the earth revolves round its axis once every 24 h and circles round the sun each year together with its moon causes many changes, both predictable and unpredictable. From the earliest times the alternation of day and night, the tides and the succession of the seasons have left their mark on the living organisms. About a century ago the physiologist Claude Bernard described the concept of homeostasis, where a series of mechanisms keeps the 'milieu intérieur' of an organism as constant as possible. They maintain a mean state around which the life functions oscillate. It is in these oscillations that we can recognize nature's rhythms: the rhythms which run synchronously with the changes of day and night, with the position of the moon with respect to the sun (the tides) and with the changing of the seasons. These rhythms have been named biological rhythms, and numerous examples can be cited. The marine crab emerges from its working place to forage at high tide, and returns to its place of concealment at low tide. Diurnal animals are active in the light and sleep in the dark, while nocturnal animals do the reverse. In autumn many birds migrate to warmer pools, to return again in springtime.

General Features

An important characteristic of these biological rhythms is that they persist under constant environmental conditions, when the plant or animals have been isolated from any direct influence from their surroundings. The rhythms are, therefore, termed endogenous, they arise within the organism and are not imposed by the environment. In conditions of constant light, plants continue to move their leaves, the marine crab becomes active every 12 h and migratory birds kept in fixed conditions of light and dark as well as temperatures still display migratory behavior at the same time as members of their species living in the wild. However, the fact that biological rhythms are endogenous does not mean that their natural period is exactly the same as that of the environment's cycles. The periods of the endogenous rhythms actually differ slightly from those in the milieu, but exogenous synchronizing cues ('Zeitgeber') adjust or 'entrain' the biological rhythms so that the organisms remain in phase with their surroundings. The daily alternation of light and dark is a powerful entrainer for most animals, but in human matters are inevitably more complex and social cues are also important.

The slight difference in period of the biological rhythms from geophysical cycles has led to their description with terms prefixed by the word 'circa', or about. Thus, one speaks of circadian (approximately a day or 24 h) and of circannual (about a year, seasonal) rhythms.

How widespread are these rhythms? With the possible exception of anucleate unicellular organisms, biological rhythms occur throughout nature, from simple uni- to highly complex multicellular organisms, including man. Another important point is that rhythms are present at every level, occurring in cells, tissues, organs and complex control systems, and arranged hierarchically. Rhythms in cells are controlled by those in the tissues. The organs are regulated neuronally and hormonally by the nervous system. Certain groups of nuclei in the hypothalamus, for example the suprachiasmatic nuclei [1], play an important part in generating various 24-hour rhythms and in adjusting them to each other. These nuclei are called pacemakers. The whole gives the impression of a very complicated clock with multitudes of interlinked cogwheels. It is these biological rhythms which allow the organism to anticipate changes in the environment: day or night, summer or winter, thus increasing its chances of survival.

History

The earliest known description of a biological rhythm is to be found in the work of Androstenes, who in about 300 BC wrote an account of Alexander The Great's campaign. He reported how the leaves of the tamarind tree move in a cycle which corresponds to the day-night alternation. This report obviously made little impression, for it was not till the 18th century that researchers, particularly the French, undertook further study of the phenomenon. It is remarkable that the field was tackled by botanists only (Pfeffer, Darwin, Kleinhoonte) until as late as the 1930s. Since then the research has broadened to include experiments using animals, primarily rats and guinea pigs, and at a later stage also human beings.

From the beginning of the investigations of human biological rhythms the endocrine system has attracted much attention. It was Pincus who 40 years ago gave the first description of a 24-hour rhythm in the urinary excretion of ketosteroids. He reported a high content in the early hours of the morning, with levels subsequently decreasing rapidly, to climb again in the course of the evening.

Rhythms, Endogenous or Induced

A major difficulty associated with studies of circadian rhythmicity in hormones in the plasma is the fact that the concentrations of free hormones are often very low. The levels of hormones in the blood are highly variable, different hormones following different patterns. It is well known that the plasma cortisol is at its highest soon after awakening and then declines; after a few hours lower secondary peaks can occur. This pattern is relatively independent of the sleeping-waking cycle, and remains detectable during sleep deprivation [2–4]. A similar pulsatile pattern is seen in plasma adrenocorticotropin (ACTH) levels [5]. As these rhythmic changes in plasma ACTH continue after adrenalectomy, they are believed to depend upon corticotropin-releasing factor release from the hypothalamus. The origin of this latter rhythm, however, is at present still unknown. By contrast, growth hormone (GH) reaches a peak concentration in the plasma in the early hours of the night. Besides this, 3–4 lower peaks can be demonstrated in the daytime. During sleep deprivation the high nighttime peak is abolished but the smaller peaks persist at intervals of about 4 h [6, 7].

In how far are hormones which fluctuate according to the sleep-waking cycle governed by endogenous rhythms? GH as stated above, is not. Prolactin probably is, with its trough level in the afternoon and gradual increase to a maximum which occurs at the end of the sleeping period. Thyroid-stimulating hormone (thyrotropin) and aldosterone show similar rhythms, but with their maxima in the early hours of the night. The rhythms mentioned so far have amplitudes of approximately 50% of the 24-hour mean. Smaller variations are seen, for example, in the plasma testosterone level, while virtually no 24-hour rhythmicity can be demonstrated in the levels of plasma luteinizing hormone and follicle-stimulating hormone.

Of the other, nonhormonal rhythms, that of the body core temperature (high in the daytime, low at night) is probably the best known. Fluctuations in performance parallel this rhythm, especially when the task involves a motor response. Experiments using memory tasks show that the rhythm of the short-term memory has its peak in the morning and that the long-term memory functions better in the late afternoon, when the body temperature also reaches its maximum. Alertness, on the other hand, is greatest around noon [8].

Thus rhythms can be demonstrated at every level. Liver cell function, kidney tubule excretion, body temperature, adrenal cortical hormones, performance levels – high and low – we are confronted with rhythmic processes, each with its own timing, its own maximum and minimum. Not the presence of a rhythm, but rather the absence of any fluctuation should make us stand up in surprise.

Development of Rhythms

Are biological rhythms congenital or do they only develop after the organism has been subject to environmental conditions for some time? Experiments in which mice were kept in constant conditions for many generations have shown that the potential of circadian rhythmicity is genetically determined [9].

For obvious reasons study of circadian rhythms in children is associated with special problems. It is not possible to obtain informed consent from an infant, and procedures like plasma sampling and urine collection from so small an individual are rather difficult, if not impossible. There are ethical considerations which hamper experiments in children that would be scientifically useful and permissible in the adult volunteer. As a result of these factors, there are much less data from children than from adults. Nevertheless, the data that are present reveal some general features about the development of the circadian control system.

In the case of humans two general rules obtain. First, the various rhythms come into being at different points in time after birth and, second, most of the rhythms appear to go through a process of maturation. After initial arrhythmicity, polyphasic fluctuations arise and then 24-hour rhythmicity gradually emerges whereas different circadian components become linked to each other. At first the circadian system is not coupled to the changes in the environment but later it is exactly synchronized to the 24-hour period of day and night.

As it is rather easy to measure body temperature in infants, the development of the temperature rhythm has been studied more extensively [10–14]. The general result is that the rhythm is absent in neonates, becoming evident around 6 months of life. From the 12th month on, the phase of the rhythm is similar to that of adults.

Data on the sleep-wakefulness pattern are obtained by either an observer or machine that measures the amount of movement. Data from Kleitman and Engelmann [15] show that during the 1st week of life the neonate spends about equal fractions of the 24 h asleep. By the 3rd week a slightly greater amount of sleep during the nighttime hours becomes apparent. This pattern becomes more marked until around 6 months of age the infant is awake most of the daytime. A more detailed analysis obtained from daily data from a single child that was demand-fed revealed that the first 4 weeks of life the sleeps are frequent, short and rather irregular during 24 h. Among 4 and 16 weeks the periods of sleep merge into a single session that appears to free run. Entrainment to the actual 24-hour time occurred around week 16 [15]. This supports the hypothesis that the rhythmic pattern of sleep is innately programmed and depends on maturational processes and that entrainment of an emerging free-running circadian rhythm occurs when the sense organs and the central nervous system allow for coupling.

There are some data from neonates that have been placed on a 4-hour feeding schedule immediately after birth and thereafter on demand-fed. In these cases, ultradian 4-hour period length rhythms alternating with irregular intervals are clearly visible and a 24-hour rhythm with 4-hour components but without free run is already present around week 6 [Siegmund, pers. commun.]. This suggests that food entrainment in the earlier phase of life might be a stronger synchronizer than the light-dark alternation. Recent findings that bright light is necessary to entrain rhythms in the adult situation might be an explanation as the level of about 2,500 lx, the intensity that is needed for synchronization, seldom will be reached in the neonates' environment.

An interesting question is whether circadian rhythms grow out of the ultradian ones. It is unknown by what mechanism ultradian components add to a circadian rhythm. One possibility is that the period gradually increases from an ultradian to a circadian value. An alternative possibility is that ultradian components drop out to produce a circadian rhythm. In this respect it is important to note that, in the adult, ultradian components as well as circadian components are present in several functions and that often these two rhythms are synchronized. Whether such a close link is due to control by the same pacemaker (for example the suprachiasmatic nuclei) is at the present time unknown.

Intermittent or episodic rather than steady-state release is the predominant mode of the anterior pituitary hormone release. A circadian rhythm exists for plasma ACTH and cortisol. Onishi et al. [16] determined serum cortisol in children ranging from 1 month to 15 years of age. They concluded that the first 5 months of life serum cortisol fluctuates in an irregular way and that a circadian rhythm appears approximately at 6 months. The main obstacle in these experiments has been the difficulty to obtain blood samples. In a later study [17], this problem was solved by taking salivary samples. Salivary cortisol shows a circadian rhythmicity around 3 months after birth, coinciding with an observed regular sleep pattern.

GH has been shown to be secreted in distinct episodes during a 24-hour day in the adult, although much emphasis has been placed on a correlation between peak secretion of GH and the onset of sleep. Sleep-related peaks in hGH appear in the first year of life, in relation to the organization of the sleep-wake cycle and the appearance of slow-wave sleep [18, 19]. Spontaneous daytime GH secretion is present in children of all ages and pubertal staging. The mentioned problem of limited sampling in infants has been exacerbated by the episodic secretion pattern which requires a high sampling rate to detect. Miller et al. [20] have demonstrated that spontaneous daytime GH secretion is present in children ranging from 2 to 18 years of age. Parker et al. [7] reported that the peak release in early sleep is maximal in pubertal and postpubertal young adult males. These groups also have a greater hGH release in the latter two thirds of sleep than do adults [7].

Although the focus is on the development of the pacemaker itself and not on the development of the mechanisms by which it controls overt functions, understanding of the latter is important for the question of how adult temporal organization (i.e. the phase relationship among the many rhythms in an individual) is formed. At the present the best system available for a study of the developmental events that lead to the appearance of overt rhythmicity is the neural regulation of the pineal gland. Other developing systems about which much is known with respect to circadian rhythmicity include the hypothalamic-pituitary-adrenal axis and the states of sleep and wakefulness.

Serum concentrations of melatonin progressively increase up to the 3rd month of life and a significant day-night difference appears thereafter [21]. In a different study Waldhauser et al. [22] described rather low levels up to 6 months. From 1 to 3 years there is an almost 10-fold increase peaking during prepuberty. The melatonin levels decrease again until at old age (70–90 years). The same low levels as in the first 6 months are reached. It is asssumed that melatonin plays a role in the internal synchronization of various circadian rhythms as well as in the transfer of synchronizing signals. In contrast to melatonin, serotonin levels exhibit a well-expressed day-night difference immediately after birth [21].

Rhythms in fluid and electrolyte state are closely determined by the fluctuation of reabsorption by the renal tubuli. Between the 6th and 16th postnatal weeks a rhythm arises in the urine production as well as in the excretion of sodium, potassium and urea. By contrast, the rhythms of phosphate and chloride excretion are not detectable till after 4 or 5 months [11]. Besides the development of 24-hour periodicity, the amplitude of rhythms also increases during maturation. Generally the mean level remains unchanged but the peaks and troughs become more marked.

In addition to these changes there are age-dependent changes in the coupling properties of circadian and higher-frequency oscillators which dominate in childhood. This is analogue to the changes in coupling between the heart frequency rhythm and that of the respiration. The variability of the heart frequency increases from child to grown-up and decreases again while aging [23].

Rhythms and Aging

Rhythms arise in the newborn. They are stable when the organism reaches adulthood and decline again as it ages. As during the development of temporal organization, the amplitude is seen to change during aging. Some oscillations become progessively flatter (potassium excretion, GH and sex hormone levels) and may completely disappear. Other rhythms remain undiminished in the elderly and even in the demented (plasma cortisol).

There are indications that the probability of internal dissociation increases with aging. This is a state in which a number of rhythms lose their synchrony and each adopts a different period. The various pacemakers lose contact with each other. A dissociation between the rhythms of body temperature and sleeping and waking occurs most frequently. Internal desynchronization normally only occurs when an experimental subject has been deprived of synchronizing signals from the environment for some considerable time, and his biological rhythms can each assume their own endogenous frequencies. The fact that the temporal organization apparently becomes looser in elderly individuals may have consequences for the organism as a whole. Studies on the life span of insects have shown that derangement of their temporal organization (for instance by subjecting them to light-dark cycles with periods greater or less than 24 h, or by repeatedly imposing phase shifts of 8 h) decreases their duration of life.

In conclusion, both basic and applied research are bit by bit revealing the mechanism which is responsible for our rhythmic functioning, as aspect of our being which is at times problematic but which we can also use to our advantage.

References

1 Meijer JH, Rietveld WJ: The neurophysiology of the suprachiasmatic circadian pacemaker in rodents. Physiol Rev 1989;69: 671–708.
2 Krieger DT, Allen W, Rizzo F, Krieger HP: Characterization of the normal temporal pattern of plasma corticosteroid levels. J Clin Endocrinol Metab 1971;32:266–280.
3 Weitzman ED, Nogeire C, Perlow M, et al: Effects of a prolonged 3-hour sleep-wake cycle on sleep stages, plasma cortisol, growth hormone and body temperature in man. J Clin Endocrinol Metab 1974;38:1018–1022.
4 Weitzman ED, Zimmerman JC, Czeisler CA, Ronda J: Cortisol secretion is inhibited during sleep in normal man. J Clin Endocrinol Metab 1983;56:352–365.
5 Krieger DT: Rhythms in CRF, ACTH, and corticosteroids; in Krieger DT (ed): Endocrine Rhythms. New York, Raven Press, 1979, p 123.
6 Quabbe H-J: Chronobiology of growth hormone secretion. Chronobiologia 1977;4:217–246.
7 Parker DC, Rossman G, Kripke DF, et al: Rhythmicity in human growth hormone concentration in plasma; in Krieger DT (ed): Endocrine Rhythms. New York, Raven Press, 1979, pp 143–173.
8 Folkard S, Monk TH: Circadian rhythms in human memory. Br J Psychol 1980;71:295–307.
9 Aschoff J: Exogenous and endogenous components in circadian rhythms. Cold Spring Harbor Symp Quant Biol 1960;25:11–28.
10 Jundell J: Ueber die nykthemeralen Temperaturschwankungen im I. Lebensjahre des Menschen. Jahrb Kinderheilkd 1904;59: 521–619.
11 Hellbrügge T: The development of circadian rhythms in infants. Cold Spring Harbor Symp Quant Biol 1960;25:311–323.
12 Hellbrügge T, Lang JE, Rutenfranz J, et al: Circadian periodicity of physiological functions in different stages of infancy and childhood. Ann NY Acad Sci 1964;117:361–373.
13 Hellbrügge T: Ontogénèse des rythmes circadiaires chez l'enfant; in Ajuriaguerra J (ed): Cycles Biologiques et Psychiatrie. Symposium Bell-Air III, Geneva, Sept 1967. Geneva, Georg Paris, Masson, 1967, pp 183–195.
14 Abe K, Fukui S: The individual development of circadian temperature rhythm in children. J Interdiscipl Cycle Res 1979;10: 227–232.
15 Kleitman N, Engelmann TG: Sleep characteristics of infants. J Appl Physiol 1953;6:269–282.
16 Onishi S, Miyazawa G, Nishimura Y, et al: Postnatal development of circadian rhythm in serum cortisol levels in children. Pediatrics 1983;72:399–404.
17 Price DA, Close GC, Fielding BA: Age of appearance of circadian rhythm in salivary cortisol values in infancy. Arch Dis Child 1983;58:454–456.
18 Finkelstein JW, Anders TF, Sachar EJ, et al: Behavioral state, sleep stage, and growth hormone levels in human infants. J Clin Endocrinol Metab 1971;32:368–371.
19 Shaywitz B, Finkelstein J, Hellman L, et al: Growth hormone in newborn infants during sleep-wake periods. Pediatrics 1971;48: 103–109.
20 Miller JD, Tannenbaum GS, Colle E, et al: Daytime pulsatile growth hormone secretion during childhood and adolescence. J Clin Endocrinol Metab 1982;55:989–994.
21 Attanasio A, Rager K, Gupta D: Ontogeny of circadian rhythmicity for melatonin, serotonin, and N-acetylserotonin in humans, J Pineal Res 1986;3:251–256.
22 Waldhauser F, Weiszenbacher G, Tatzer E, et al: Alterations in nocturnal serum melatonin levels in humans with growth and aging. J Clin Endocrinol Metab 1988;66:648–652.
23 Rensing L: Chronobiologie des Alterns: Veränderungen der zeitlich-periodischen Ordnung. Z Gerontol 1989;22:73–78.

Prof. W.J. Rietveld
Department of Physiology
University of Leiden
PO Box 9604
NL–2300 RC Leiden (The Netherlands)

Horm Res 1990;33:58–68

0301–0163/90/0334–0058$2.75/0

Somatomedins on the Move[1]

J.L. Van den Brande

Wilhelmina Children's Hospital, University Department of Paediatrics, Utrecht, The Netherlands

Mister Chairman, Ladies and Gentlemen,

I am greatly honored and pleased to receive the Andrea Prader prize from the society which I treasure so much. It is an award, not only for myself, but even more so for all those whom I had a chance to work with.

I must confess, however, that the holidays which I planned to have in Greece before the ESPE Summer School and this meeting, though physically realized, were filled with thinking, writing, faxing and phoning. I am therefore glad that you never will give me this prize again. Andrea Prader is the name attached to this prize and rightly so. He is the father of pediatric endocrinology in Europe. This position is also exemplified by the image I had of him during the first years of my attendance to our meetings, sitting in the middle of the front row at our meetings. Fortunately I got a bit closer to him in later years and I want to thank Andrea for the impact he has had on me and many of us. I would like to situate myself in the family of pediatric endocrinologists. This is shown in a small section of the family tree (fig. 1). I was adopted by Judson van Wyk, who himself was an adopted son of Lawson Wilkins. Thus, I am a grandson in straight line of this grand man, founder of pediatric endocrinology.

However, another person was instrumental in this adoption: Henk Visser. I want to reveal here how I got into pediatric endocrinology. When finishing my pediatric training with the late Prof. Hooft in Gent, Belgium, in 1964, I had the idea that cancer should be the consequence of losing the function of a protein and, since this should be a structural loss, I thought perhaps one could, by immunological means, isolate this protein and get grip on carcinogenesis. I had come to know Henk well enough in the previous years that, when I was searching for a center where to start working on this problem, I contacted him and he invited me to come to Groningen and discuss the matter. Henk cleared his agenda and during two, three full days we walked in the wide polders and along the canals. We discussed the feasability at that time of solving the specific problem I had in mind and in which areas, in Henk's opinion, things were moving. This brought us to endocrinology: regulation of bodily functions, growth and development. You are not surprised. These were problems right in the heart of pediatrics, of which so little was understood. I became enthusiastic about the prospect of moving into that field and was converted to pediatric endocrinology. And I went, with my family, to Chapel Hill. A tremendous change of climate in all meanings of the word. There I was to work with Jud van Wyk whose efforts were all directed towards turning me from the mere clinician I was into a scientist. I have been caught by the virus since.

To these men I owe much of my scientific development and I will always remain grateful to them.

In later years I have had an opportunity to adopt quite a number of scientific sons and daughters as fellows myself and a few of them as permanent collaborators, and I want to stress that their work is what I am honored for today. I should especially mention the collatoration with John Sussenbach and his group which has proved to be so fruitful. We all have had support from a series of technicians, whom I want to mention here as well (table 1).

Finally, I want to thank my wife for having accepted leaving Belgium in the sixties though it meant never returning there because of lack of opportunities, and

[1] Andrea Prader Lecture held in Jerusalem on November 1, 1989, on the occasion of the remittance of the Andrea Prader Prize at the Third Joint Meeting of the European Society for Pediatric Endocrinology and the Lawson Wilkins Pediatric Endocrine Society.

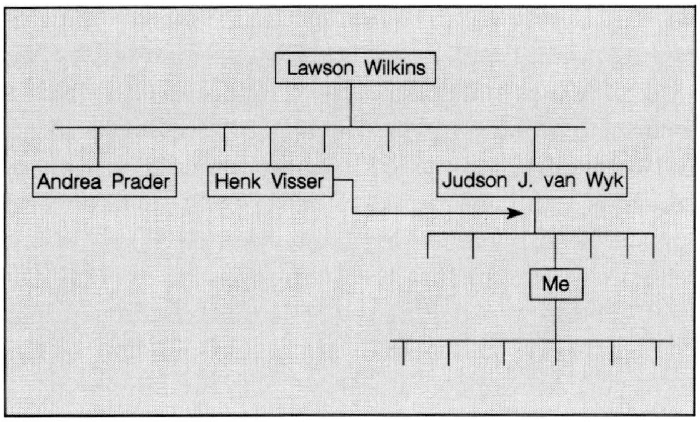

1

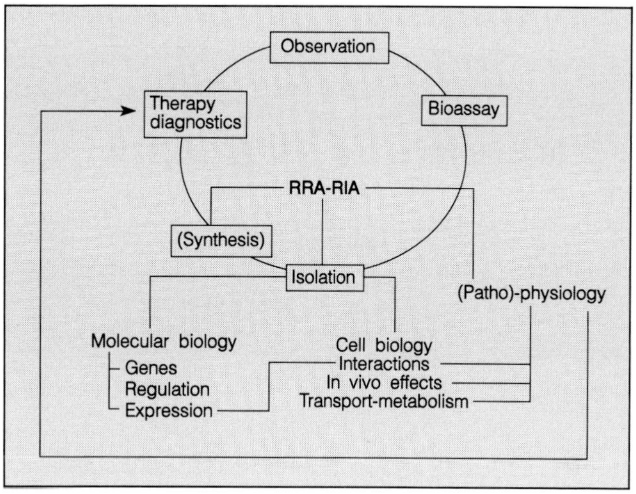

2

Fig. 1. Extract from the pediatric endocrinology family tree.

Fig. 2. Classical cycle of scientific progress in endocrinology. RRA = Radioreceptor assay; RIA = radioimmunoassay.

my children for having endured so well the hardship of having a father much engaged in other things than the family.

Ladies and gentlemen, in this presentation, I would like to follow with you how endocrinology is on the move and I will do so using the somatomedin or insulin-like growth factor (IGF) research as a model.

The classical cycle of gathering knowledge in endocrinology involves an observation, setting up a bioassay, isolation of the active substance and studying its physiology and pathophysiology, produce the substance for application, which will lead to new observations (fig. 2). Cycles involving different processes and substances are of course interwoven and generate each other. Such the discovery of sulfation factor was the consequence of an event in the growth hormone (GH) cycle (fig. 3). When Daughaday was trying to establish a bioassay for GH, with his fellow Salmon, they made the observation that, though GH induced sulfate incorporation in cartilage of an injected animal, it did not do so when incubated in vitro with this tissue. They had the brightness not to blame the technique but to suggest that GH perhaps induced another factor which in turn did the job. They tested this hypothesis with a classical experiment showing that GH injection causes the inactive serum of a hypox rat to become active in stimulating in vitro incorporation in cartilage. At the same time the cartilage of the injected animal also becomes active but with some delay [1]. Based on these observations they generated the hypothesis that a 'sulfation factor' was instrumental in transmitting the GH signal to the tissues. At the same time they generated a bioassay, as an instrument for further study.

Table 1. Collaborators, past and present (*)

Fellows and staff	Technicians
Department of Pediatrics, University of Rotterdam and Utrecht	
M.V.L. Du Caju	M. Vos
T. Von Ruhe Zurcher	C.M.Hoogerbrugge*
S. van Buul-Offers*	T. Huyser
J.M. Wit*	Y. Schouwstra
D.A. Price	A.M. Korteland-van Male
J.J. Bongers-Schokking	L. Dumoleijn
J.A. Bakker	P.L.M. van der Klundert
A.J.M. Smeets	M. Feylbrief
M. Jansen*	J. Branger
H. Delemarre	R. Reijnen-Gresnigt*
R. Odink	T. de Poorter
J. Stokvis	R. Bloemen*
J. Jansen	E. Bonte*
C. Logghe*	E. Ram*
B. Bakker*	
Collaboration with the Department of Physiological Chemistry, University of Utrecht	
J. Sussenbach*, chairman	R. van Schaik*
E. Holthuyzen*	
Pediatrics and Physiological Chemistry	
S.L. Oei*	
D. Meinsma*	
L. de Moor*	
W. Bovenberg*	

When I chose this subject as a fellow of Jud Van Wyk, the first we did was to try and isolate this factor, with little success, however, due to the lack of a more easy and precise bioassay. Back in Holland, with my fellow Marc Du Caju, we developed the porcine cartilage bioassay

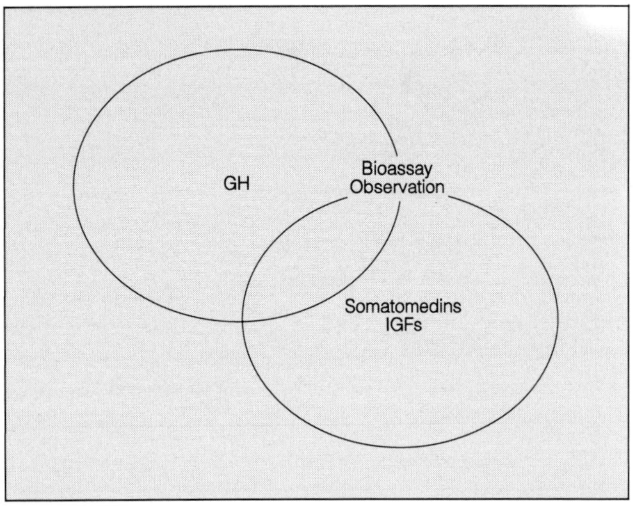

Fig. 3. Generation of the somatomedin cycle in the GH cycle.

[2]. By the way I treasure this choice, since I have great pity with the pig, which in my opinion is a real drama of nature. You know that the pig, next to the primates, has the furthest evolution of its proteins. Why then is it so backward? Davison and Dobbing implicitly gave the answer in their study of the rate of brain growth of a series of species. All species exhibit nice sharp peaks, contrasting with the flattened peak of the poor pig! A mutation or an experimental flaw? Certainly a drama. But back to serious business. This bioassay allowed us and others to study the physiology of the sulfation factor [3], later called somatomedins, and to proceed with isolation [4].

Van Wyk and his collaborators wanted to have the pure substance. This work was facilitated by the finding of Hall and himself that somatomedin activity, associated with a large molecular fraction, became relatively small upon acidification [5]. One of his major goals was to raise antibodies, and with his collaborators he did that with success, allowing many laboratories all over the world to proceed more rapidly and accurately with their studies.

Rinderknecht and Humbel [6, 7] described the structure of two somatomedins, renamed IGF-I and IGF-II, a logical designation from the background of their work with Rudi Froesch on nonsuppressible insulin-like activity, but also from the structural analogy with insulin which I shall discuss later. When also multiplication stimulating activity (MSA) was shown to be an IGF these three lines of investigation converged. The availability of pure IGFs from Humbel's laboratory again boosted IGF research. In the meantime the laboratories of Jürgen Zapf and Ray Hintz, closely followed by the work of Moses and collaborators, independently discovered the binding proteins (BPs) [8–10] present in serum in two main fractions of 150 and 40 kilodaltons (kDa), which would later prove to have such an important impact on IGF action, and D'Ercole et al. in van Wyk's laboratory showed that IGFs were present in many tissues, thereby broadening the endocrine concept to that of a paracrine and perhaps autocrine function of the IGFs [11]. My collaborator Sylvia van Buul in this same period obtained preliminary evidence for effects of somatomedins in vivo, similar to those of GH, shown in figure 4 [12].

The next era is that of molecular biology which started when my fellow Maarten Jansen supported by John Sussenbach isolated and sequenced cDNAs encoding the IGFs, thereby showing that they are synthesized as preprohormones [13]. With the discovery that IGF-I had its own receptor separate from the insulin receptor [14], the cloning of the cDNAs for the BPs [15–17], the discovery of interactions of IGFs with other growth factors [18], and the availability of these substances in pure state allowing in vivo experimentation and application in diagnostics and therapy, the cycle is complete. However, many questions remain and in the next part of my talk I want to focus on some of them. At the end, I will try and convince you that nature has given us a flexible system to control growth and metabolism which adjusts itself to the needs in time and place and to the external circumstances.

Let's begin with the origin: the genes.

The human IGF-I gene is a very large gene of which unfortunately we know still very little (fig. 5) [19, 20]. It comprises at least 5 exons. By alternative splicing 2 different types of mRNAs can be generated which differ partially in the region encoding the E domain of the prohormone. Thus, IGF-Ia and -b can be generated. The mature IGF resulting from the removal of the E domain is identical. The difference in function of these 2 mRNAs and prohormones has not yet been resolved. Much work remains to be done on this gene since we do not know its full organization. In the rat a much more complex organization has been found [19]. It is, therefore, likely that the human gene still hides a lot of secrets!

The human IGF-II gene is better known. The work of Elly Holthuizen in Sussenbach's laboratory has greatly contributed to this [20]. It is an unusual gene since it contains 4 promoters allowing the generation of many

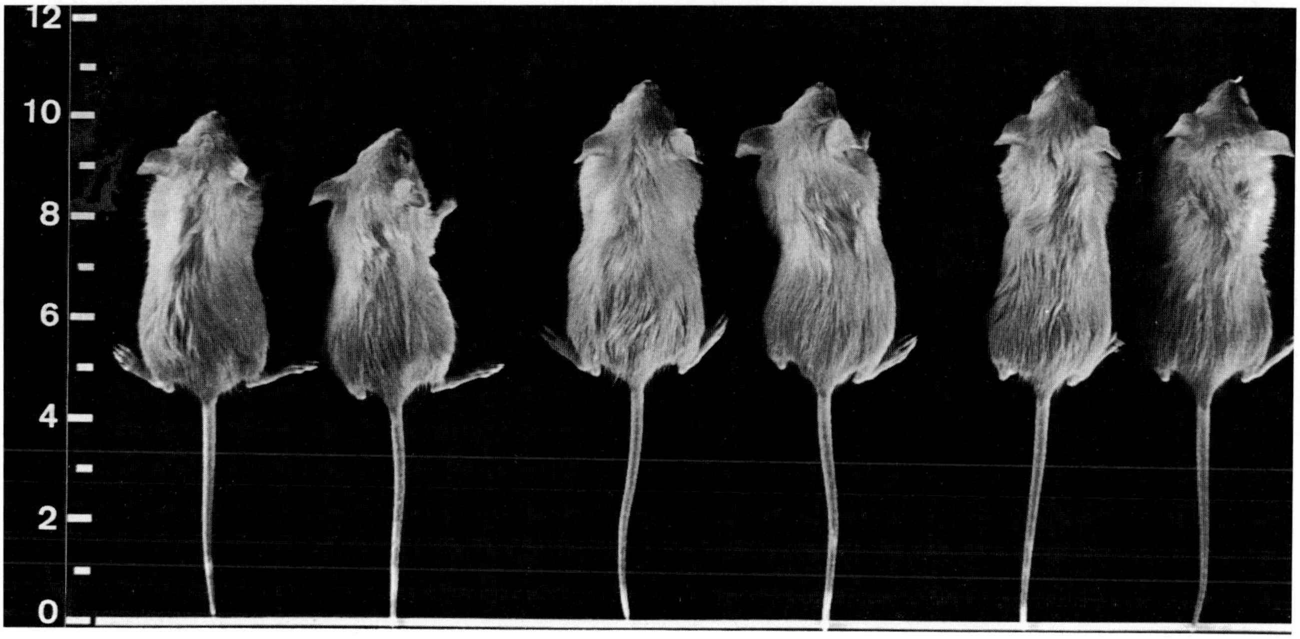

Fig. 4. Aspect of dwarf mice, treated for 4 weeks with phosphate buffer (2, animals on the left), a semipurified somatomedin preparation (6 U/day; 2 animals in the middle) and hGH (8.3 µg/day; 2 animals on the right). The scale on the ordinate is in centimeters. From van Buul-Offers et al. [12].

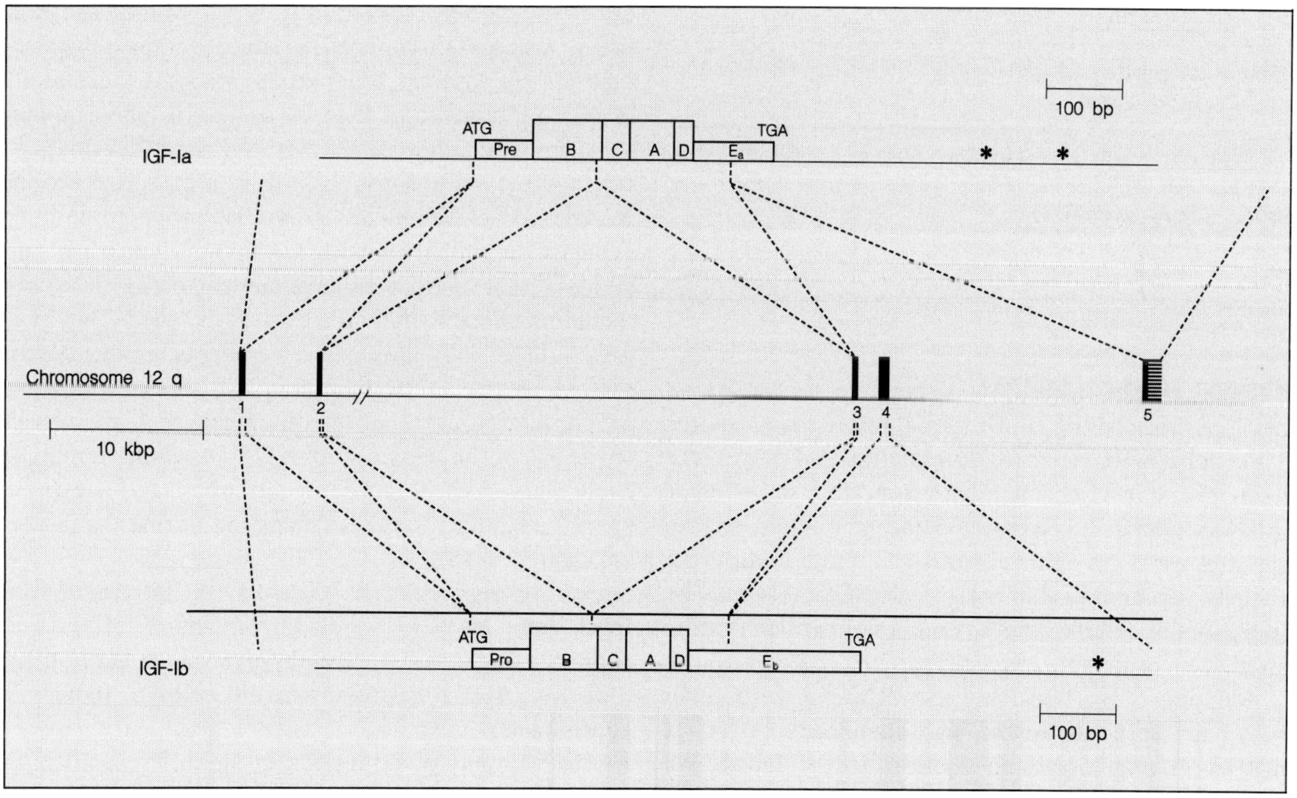

Fig. 5. Map of the human IGF-I gene and mRNAs. The coding regions are indicated by blocks [13, 19, 20]. bp = Basepair.

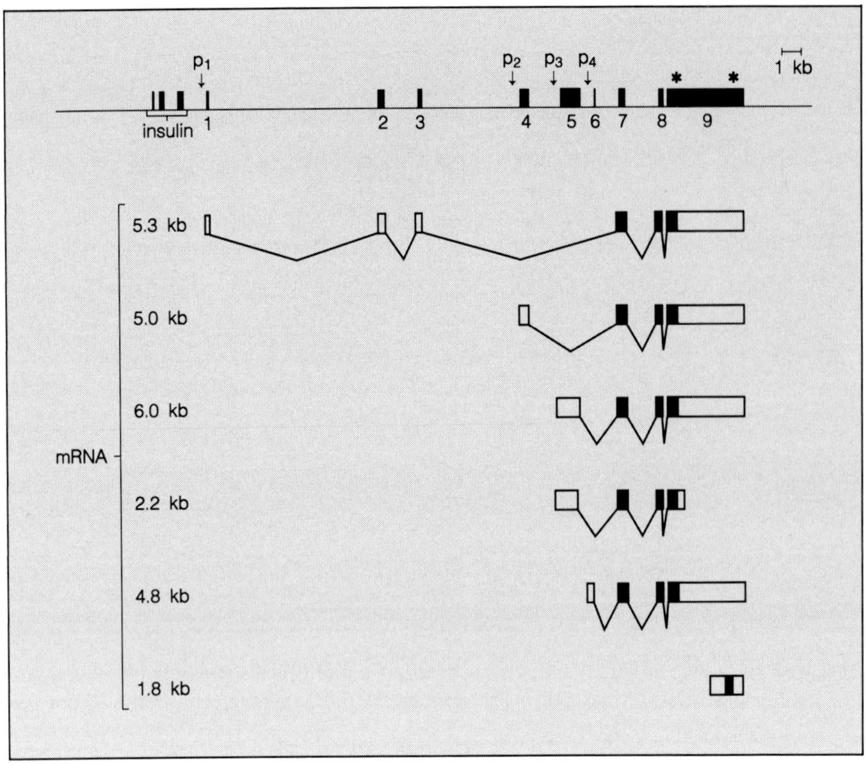

Fig. 6. Map of the human IGF-II gene and mRNAs. In the mRNAs the coding regions are indicated by black boxes. Polyadenylation sites are indicated by asterisks [19–21]. kb = Kilobase.

Table 2. Human IGF-BPs

Type	Originally isolated from	Plasma		Hormonal influence	
		150 kDa	40 kDa	insulin	GH
BP1	amniotic fluid	–	+	↑	↓
BP2	fetal liver				
BP3	plasma	+	+	–	↑

different species of mRNA, depending on the exons being transcribed (fig. 6). Of special interest is promotor 1 which becomes active in late gestation and only in the liver. The IGF-I gene generates a number of different mRNAs as well. It is a major challenge to elucidate the significance of the numerous mRNAs found in different tissues. Variation in stability and translatability may be important in determining the amount of peptide synthesized at any time in each cell type.

Mapping of the IGF-I and -II mRNAs in human tissues has clearly shown the high abundance of IGF-II mRNA in fetal tissues in contrast to the low IGF-I mRNAs [22]. When in situ hybridization studies were done on human fetal tissues Han and collaborators [23] did a remarkable observation: only cells of mesenchymal origin, thus mainly connective tissue, contained the message [23]. In contrast, other groups studying the amount of IGF protein found it to be present in parenchymal cells, amongst others [24]. This suggested that the peptide had moved towards these cells: a paracrine action. Recently, Paul Schofield's group has proposed another explanation which fits with the finding that when one studies tissues earlier in development many cell types including parenchymal cells express the gene. They suggest, without excluding paracrine effects in later stages, that expression is most pronounced in cells which are differentiating, and that the differences seen are related to the age of the fetuses studies [25]. A similar temporal change has been observed in rats where IGF-I expression, though low, increases manyfold during the period of organogenesis.

All this suggests that IGFs play an important role from quite early in development, actually IGF-II is expressed in the 4–8 stage embryo development in mice [26]. Also IGF-I is present in the embryo, though it appears later.

The mapping of expression of the different mRNAs during development will keep many a scientist at work for a long time!

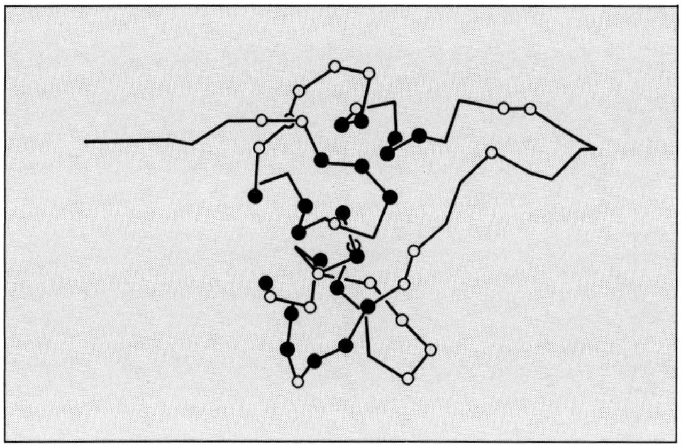

Fig. 7. Model of IGF-II structure derived from insulin coordinates. ● = Residues conserved between insulin IGF-I and -II; ○ = residues identical in human IGF-I and -II, but not in insulin. From A. Honegger [27].

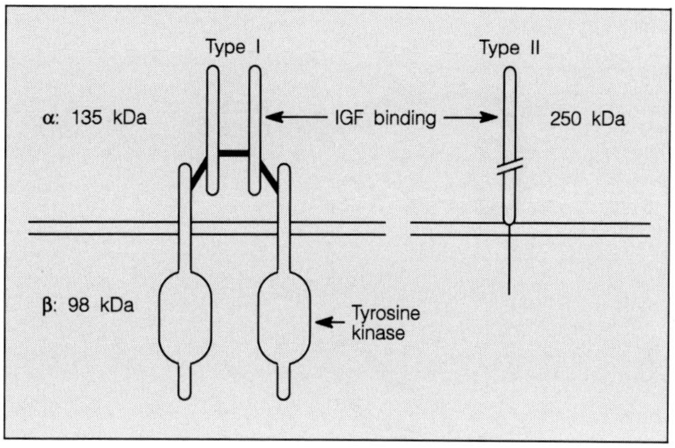

Fig. 8. The IGF receptors.

Let me now turn to the protein. The IGFs, insulin and relaxin are structurally quite related. This is nicely demonstrated for the IGFs and insulin in figure 7, drawn by Annemarie Honnegger [27] and based on the tertiary structure model proposed by Blundell et al. [28] (fig. 9). In the core of the molecule the main chains coincide; only in the periphery do they diverge, accounting for the difference in function. How these differences influence interactions with other molecules can best be studied with natural or synthetic variants. But before going into that I must briefly present the ligands with which IGFs can bind.

First, the BPs. In view of the nice review which Ray Hintz gave on this topic [29, this volume], I shall limit myself to reminding you of the 3 BPs of which the full sequence is known (table 2). BP1 or BP28 was originally isolated from amniotic fluid by Drop et al. [30] and Chochinov et al. [31]. Antibodies against this molecule cross-react with the 40-kDa fraction in serum. BP2 is the so-called fetal form. Its sequence in the human was recently reported by Binkert et al. [17]. It is most likely that this or a related molecule is also measured in serum in the 40-kDa range with the presently available antibodies against BP28. BP3 or BP53 is the IGF-binding subunit of the 150-kDa complex isolated by Baxter and Martin [32]. All these BPs have an important amino acid homology. The cysteine residues are fully preserved. An important difference is the extensive glycosylation of BP3.

By western blotting more species are found which may represent different molecules or modifications of these 3 types [33]. Whereas the production of BP1 appears to be ubiquitous, BP3 production is limited to fibroblasts, endothelial cells, anterior pituitary, postnatal liver and possibly endothelial cells. [32].

Next, let me remind you of the third party in this game: the IGF receptors (fig. 8). Type I binds both IGF-I and -II very well, I better than II; type II preferentially binds IGF-II but exhibits some cross-reactivity with IGF-I as well. Insulin binds weakly to the type I receptor and not to type II [34]. These molecules are vastly different. The function of the type II receptor, which is an evolutionary product of a mannose 6-phosphate receptor [35], is unknown. The type I receptor transduces the signal leading to mitosis but also appears to induce a metabolic event such as glycogen synthesis in the diaphragm [36].

Now that we have the players into place, let us examine where the interaction between IGF molecules, receptors and BPs takes place. This is not only an exciting intellectual exercise but also relevant if we want to understand the physiology and if we are to manipulate the properties of the molecules for specific applications. From the model proposed by Blundell et al. [28], Honegger [27] has proposed that the interaction with the insulin receptor takes place in the region of the B domain (B25–26) which, in the tertiary structure model of the IGFs, is located between the C and D domains (fig. 9). The interaction expands across the molecule to involve neighboring regions of the A and B domains (B12, 13, 16 and A19, 21). The interaction with the type I receptor is supposed to take place in the same region. Honegger [27] proposed that both the C and D domains present in IGFs

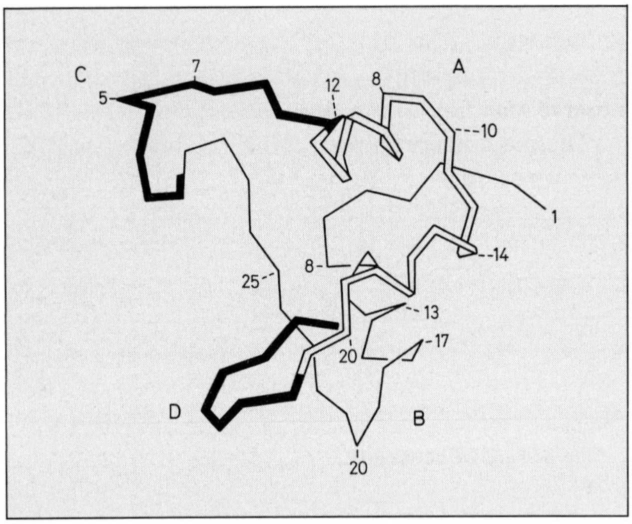

9

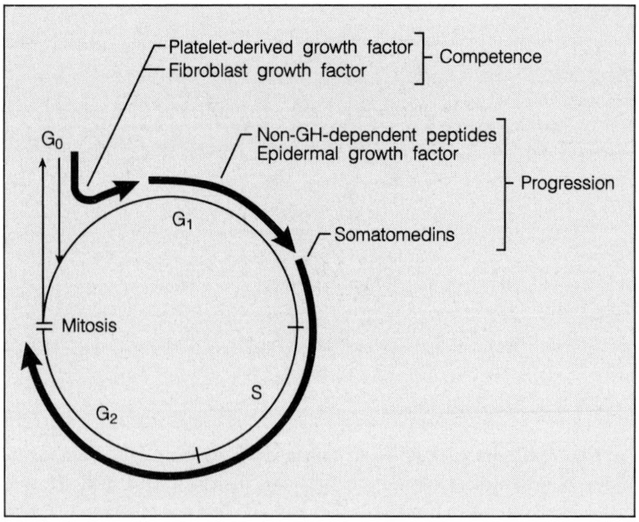

10

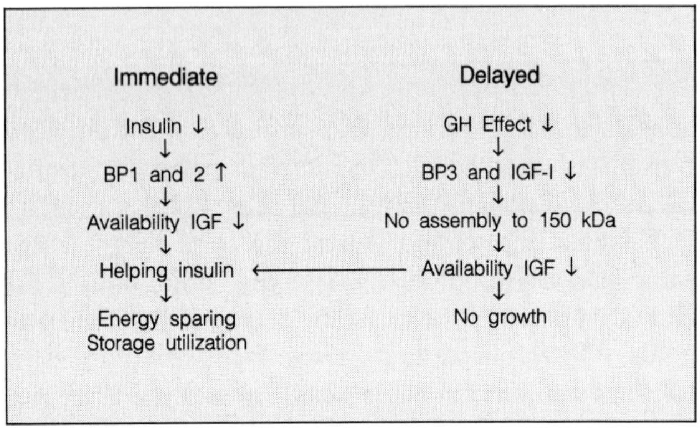

11

Fig. 9. Tertiary structure of IGF-I, looked at from the opposite side as at IGF-II in figure 7. —— = B domain; ■ = C and D domains; ▭ = A domain. The numbers indicate amino acid positions within the domain. From A. Honegger [27].

Fig. 10. A model of the involvement of growth factors in different phases of the cell cycle in fibroblasts as proposed by Stiles et al. [18].

Fig. 11. A proposal for a biphasic response (imediate and late) of IGFs and IGF-BPs to a low nutrient intake, adjusting the flow of energy and reducing growth.

and not in insulin, by their bulkiness, hamper interaction of IGFs with the insulin receptor. However, recent studies by Cascieri's group [36, 37] have shown that removal of the C domain does not influence insulin receptor binding. On the other hand, removal of the D domain increases interaction of IGF-I with the insulin receptor [38], thus showing that this domain restricts interaction with the insulin receptor. However, in view of the limited increase (2-fold), other regions of the molecule must be involved in this interaction as well.

An IGF-I clipped between C7 and C8 studied by us has a 30-fold reduction of reactivity with the type I receptor [38] and also Cascieri et al. [36] have shown that removal of the C domain yields a similar result. Therefore, in contrast with the insulin receptor interaction, the C domain is necessary for interaction with the type I receptor. Consensus seems to be reached that the area which was suggested by Honegger [27] to be involved with the interaction with the type II receptor [24] is indeed located in the region of the A domain involving amino acids A8–10, but also involving part of the B domain [40, 41]. On the other hand, a more detailed knowledge on the interaction site with the BPs has evolved since the initial proposals by Honegger [27], who located it in the region of the molecule involving B11–23 as well as the D domain. Carlsson-Skwirut et al. [42] in Sara's group have demonstrated that a naturally occurring variant of IGF-I, in which the 3 amino acids at the N terminus are clipped [des-(1-3)-IGF-I], causes a 100-fold reduction of binding to BP1. In addition, biological activity in vitro is enhanced [42]. I shall discuss the possible consequence of these findings later. For the time being it is enough to say that the N-terminal region, at least of IGF-I, is important for binding to BP. In addition, Cascieri et al. [36] and Bayne et al. [41] have demonstrated using IGF-I mutants that the whole region around the N terminus is involved in the interaction with all types of BP present in serum.

Until a few years ago there was consensus that the BPs are capable of inhibiting interaction of IGFs with their receptors [30, 43, 44]. However, when purifying BPs from amniotic fluid, Busby et al. [45] have found two fractions: one stimulating and one inhibiting the effect of IGF-I on porcine aortic smooth muscle cells. These fractions also differed in their capacity to adhere to the cell surface, the one which stimulates adheres better than the one which inhibits. This characteristic may be linked to the difference in interference with the IGF-I action. De Mellow and Baxter [46] have recently shown that BP3, which, when incubated simultaneously with IGF-I has an inhibiting effect, has a potentiating effect when neonatal skin fibroblasts are preincubated with it before exposing them to IGF-I. Therefore, either processing of the molecules or tissue- or cell-specific characteristics can dramatically modulate these interactions. I have not added the variation in number and affinity of receptors in different cell types into this complicated equation. Clearly, looking at the chain of events from IGF gene transcription, and translation, interaction with BPs and receptors, a tremendous variability is potentially present at all these levels to adjust the final cellular responses.

However, IGFs by themselves seem to be poor mitogens. IGF-I, which is the only one to have been studied in this respect in some detail, needs other growth factors for a full effect. The first comprehensive studies approaching this aspect were done by Stiles et al. [18] in fibroblasts (fig. 10). Other factors are responsible for bringing cells from G_0 into G_1, thus rendering them competent. Next IGF-I stimulates the cell to progress through the rest of the cycle. However, it is well known that growth factors may assume different roles in different cell types. This may also apply to the IGFs. In cartilage IGF may be the competence factor and epidermal growth factor, the progression factor, as was recently suggested by Hiraki et al. [47]. It appears therefore that, if we want to gain true insight in local growth regulation, it will be necessary to take one tissue at a specific time in development and study production and interaction of all these substances. Limitation of the place of action, but expansion of the number of actors should be the devise.

Next, I want to make some considerations on the endocrinological physiology of the IGFs. There is no doubt that IGFs can stimulate body growth though there may be quantitative differences in the effects on specific organs. Our preliminary results on mice have been amply confirmed using more pure and later recombinant IGF-I by Schoenle et al. [48] and ourselves [49]. Preliminary short-term experiments in man have shown that IGF-I, besides inducing insulin-like effects, elicits a response similar to that seen with GH. Thus, Guler et al. [50] demonstrated increased glomerular filtration rate and renal plasma flow in 2 normal adults as one may observe after GH administration.

The main issue, however, with regards to the endocrine versus para- or autocrine role of IGFs is: how much of this enormous store of IGF which is present in blood is available to tissues, to add to the effect of locally produced IGFs. Though we do not know the answer in quantitative terms, some considerations of a qualitative nature may be helpful. It appears from studies on IGF-I levels in plasma in physiology and in pathological conditions that they adjust themselves to the suitability of the condition for growth [3, 19]. This is most clearly illustrated by its lowering in malnutrition. This adjustment is partially mediated by an adjustment of GH secretion or effect [51]. Without neglecting the changes in responsiveness in the target tissues or the effect of other hormones or growth factors [52, 53], a change in IGF-I levels appears to help adjusting growth to the environmental and endogenous conditions.

Modulation of IGF-BPs may add to adjust the IGF availability to tissues. It has been shown extensively that the 150-kDa complex which contains most of the circulating IGF is heavily GH-dependent and shows very little fluctuation through the day as measured by its main constituent BP3. On fasting it has recently been shown in the pig to be reduced drastically [54]. It is quite likely, but needs further documentation, that such also happens in man. Since it has been shown that BP3 production is stimulated by IGF-I [55], this reduction may be the consequence of the drop in IGF-I, a reduced GH effect, or both. In contrast, the BPs present in the 40-kDa complex are fluctuating widely through the day and are lowered by insulin [56–58].

The regulation of BPs by GH and insulin should bring back into focus the role of these hormones in adjusting metabolism to the nutritional status, as studied in the sixties by the late David Rabinovitch (Rabin) and others [59]. GH for energy sparing during fasting; insulin for utilization and storage when feasting. The BPs appear to reinforce the hormonal adjustment to fasting or abundance of nutrients. In case of fasting, BP1 and 2 by an acute increase limit the tissue availability of IGFs, thereby supporting the low insulin levels in the sparing of energy. BP3, and of course the reduced IGF-I itself, as a slower mechanism reduces the amount of IGF present in the blood, also assisting low insulin, but at the same time reducing growth (fig. 11). If conditions are favorable the shifts move in the opposite direction allowing growth and, in case of abundance of nutrients and thus high insulin, storage.

As to IGF-II, in man it is present in blood from early in development into adulthood. Its concentration is barely influenced by hormones and external factors such as acute fasting [60]. It appears to have a basal role in functions which must be continuously secured. However, the quantity of it available to tissues may be also reduced in fasting or malnutrition by the changing concentrations of the BPs. These endocrine adjustments should be added to local events, where GH and insulin will have direct effects on IGFs and BPs. In addition, changes of GH concentration or effect, by virtue of its role in clonal differentiation and consequently local IGF-I production and clonal expansion [61, 62], will have long-lasting consequences. In specific areas like the brain, but perhaps elsewhere as well, proteolytic cleavage of IGF-I to des-(1-3)-IGF-I (truncated IGF-I) may selectively occur to reduce inhibition by local BPs and attain full but, in view of rapid metabolization, short-range growth-promoting effects in time and place. Finally, local processing of BPs or a separate regulation in particular tissues may annihilate their inhibitory effect on IGF availability or even yield increased bioavailability in order to protect certain tissues.

Mr. Chairman,

while we are completing the cycle of the IGFs, the GH cycle from which it originated has now been more fully completed. Thanks to the recombinant DNA technology its availability is unlimited and its use in non-GH-deficient conditions is being explored. From the mere existence of a dose-response relationship we can expect always to obtain some increment in growth velocity with hGH. However, shall we always influence final stature without having to use an extreme pharmacological dosage? Many groups are presently investigating this question. Wit, in our group, has now obtained results on growth and bone age in short children with apparently normal hGH secretion (as evaluated by different methods) after 2 years of hGH treatment. They clearly show that while the standard deviation score for height is improved but slowly declines in most of the children, height for bone age is insignificantly influenced [63]. It seems, therefore, that if we are to improve final height of these children we also shall have to learn more about skeletal maturation and how to influence it.

This brings me to the end of my talk. Our motive for studying growth is perhaps curiosity but, more importantly, aimed at improving the prospects of children with growth disturbances. We have reached some results with hGH, but we will have to learn much more about IGFs and their BPs, and all other hormones and growth factors their action is interwoven with. We still have to learn almost everything about the relationship of endocrine and paracrine components in growth regulation. We have yet to understand the details of signal transduction and the involvement of many intracellular substances and structures. We have to learn about growth gradients through the body and how they are produced and maintained. All this we will need to study if we are to understand, not only overall growth, but also local growth and its dysregulation. This also involves derangements leading to congenital malformations. Since we know that growth factors are involved in these early processes, in my view the study of normal and abnormal embryonic development has become a working field for pediatric endocrinology as well.

In order to accomplish this we shall need to attract many young bright people. Equally important will be to have intensive and open communication amongst scientists. In addition to formal and informal direct exchange in and outside meetings, I envision a totally new type of journal as an instrument for this communication. A computer journal, containing the raw data besides the present-day format of presentation, not controlled by referees but with invited discussants and open to public discussion. This should bring the responsibility fully back to the authors, where it belongs. They will be wise enough to ask senior colleagues to criticize their data and interpretation before submitting. Free endocrinology.

Most importantly, we will have to join forces with groups working with the same aims but on different aspects. We should tune in with them, and they with us.

As a young man I once put a songbook together used in youth movements. The introduction I wrote for it seems rather applicable to our situation, at least I hope it does. It said:

We had put out the fire
And now in a broad front
We walked over a wide sandy road
An old song supported the rhythm of our muffled steps.

Suddenly we stood still and listened:
In the night far away our song was carried on.

Then we heard many voices from all around joining
At last wholly filling the silence

We looked at each other and smiled:
Friends had heard us.

I have said.

Acknowledgements

My sincere thanks go to my secretary, Mrs. K. Keulers and Mr. J.H. Glasstra van Loon and collaborators of the Central Audio-Visual Centre, University of Utrecht, for their skillful work in preparing the manuscript and figures.

References

1 Salmon WD Jr, Daughaday WH: A hormonally controlled serum factor which stimulates sulfate incorporation by cartilage in vitro. J Lab Clin Med 1957;49:825–836.
2 Van den Brande JVL, Du Caju MVL: An improved technique for measuring somatomedin activity in vitro. Acta Endocrinol 1974;75:233–242.
3 Van den Brande JL: Plasma somatomedin: Clinical observations; in Pecile A, Müller EE (eds): Growth Hormone and Related Peptides. Int Congr Ser Amsterdam, Excerpta Medica, 1976, vol 381, pp 271–285.
4 Van den Brande JL, Hoogerbrugge CM: Plasma fractions containing somatomedin activity. Part I: Extraction and partial characterization; in Pecile A, Müller EE (eds): Growth Hormone and Other Biologically Active Peptides. Amsterdam, Excerpta Medica, 1980, pp 81–102.
5 van Wyk JJ, Hall K, Weaver RP: Partial purification of sulphation factor and thymidine factor from plasma. Biochim Biophys Acta 1969;192:560–562.
6 Rinderknecht E, Humbel RE: The amino acid sequence of human insulin-like growth factor I and its structural homology with proinsulin. J Biol Chem 1978;253:2769–2776.
7 Rinderknecht E, Humbel RE: Primary structure of human insulin-like growth factor II. FEBS Lett 1978;89:283–286.
8 Zapf J, Waldvogel M, Froesch ER: Binding of nonsuppressible insulin-like activity to human serum: Evidence for a carrier protein. Arch Biochem Biophys 1975;168:638–645.
9 Hintz RL, Liu F: Demonstration of specific plasma protein binding sites for somatomedin. J Clin Endocrinol Metab 1977; 45:988–995.
10 Moses AC, Nissley SP, Cohen KL, Pechler RMM: Specific binding of a somatomedin-like polypeptide in rat serum depends on growth hormone. Nature 1976;263:137–140.
11 D'Ercole AJ, Appelwhite GT, Underwood LE: Evidence that somatomedin is synthesized by multiple tissues in the fetus. Dev Biol 1980;75:315–328.
12 van Buul-Offers SC, Van den Brande JL, Hoogerbrugge CM, et al: Effects of growth hormone and peptide fractions containing somatomedin activity on growth and cartilage metabolism of Snell dwarf mice. Acta Endocrinol 1979;92:242–257.
13 Jansen M, Van Schaik FMA, Ricker AT, et al: Sequence of a cDNA encoding human insulin like growth factor I precursor. Nature 1983;306:609–611.
14 Marshall RN, Underwood LE, Voina SJ, et al: Characterization of the insulin and somatomedin-C receptors in human placental cell membranes. J Clin Endocrinol Metab 1974;39:283–292.
15 Brinkman A, Groffen C, Kortleve DJ, et al: Isolation and characterization of a cDNA encoding the low molecular weight insulin-like growth factor binding protein (IPB-I). EMBO J 1988;7: 2417–2423.
16 Wood WI, Cachianes G, Henzel WJ, et al: Cloning and expression of the growth hormone-dependent insulin-like growth factor-binding protein. Molec Endocrinol 1988;2:1176–1185.
17 Binkert, C, Landwehr J, Mary J-L, et al: Cloning sequence analysis and expression of a cDNA encoding a novel insulin-like growth factor binding protein (IGFBP-2). EMBO J 1989;8: 2497–2502.
18 Stiles CD, Capone GT, Scher CD, et al: Dual control of cell growth by somatomedins and platelet-derived growth factor. Proc natl Acad Sci USA 1979;76:1279–1283.
19 Daughaday WH, Rotwein P: Insulin-like growth factors I and II: Peptide, messenger ribonucleic acid and gene structures, serum, and tissue concentrations. Endocr Rev 1989;10:68–91.
20 Sussenbach JS: The gene structure of the insulin-like growth factor family. Prog Growth Factor Res 1989;1:33–48.
21 De Pagter-Holthuizen P, Jansen M, Van der Kammen RA, et al: Differential expression of the human insulin-like growth factor II gene: Characterization of the IGF-II mRNAs and mRNA encoding a putative IGF-II-associated protein. Biochim Biophys Acta 1988;950:282–295.
22 Han VKM, Lund PK, Lee DC, D'Ercole AJ: Expression of somatomedin/insulin-like growth factor messenger ribonucleic acids in the human fetus: Identification, characterization, and tissue distribution. J Clin Endocrinol Metab 1988;66:422–429.
23 Han VKM, D'Ercole AJ, Lund PK: Cellular localization of somatomedin (insulin-like growth factor) messenger RNA in the human fetus. Sci Rep 1987;23:193–197.
24 Hill DJ, Clemmons DR, Wilson S, et al: Immunological distribution of one form of insulin-like growth factor (IGF-binding protein and IGF peptides in human fetal tissues. J Mol Endocrinol 1989;2:31–38.
25 Brice AL, Cheetham JE, Bolton VN, et al: Temporal changes in the expression of the insulin-like growth factor II gene associated with tissue maturation in the human fetus. Development 1989; 106:543–554.
26 Rappolee DA, Werb Z, Pedersen R, Sturm K, Schultz GA: Expression of genes for growth factors and their receptors in preimplantation mouse embryos. Cell Diff Dev 1989;27(suppl): 512.
27 Honegger A: Bovine Insulin-Like Growth Factors I and II: Purification from Foetal Calf Serum and from Adult Bovine Serum, Immunological and Receptor Crossreactivities, Amino Acid Sequences and Tertiary Structure Models; Ph.D. thesis, University of Zürich, 1985.
28 Blundell TL, Bedarkar S, Rinderknecht E, Humbel RE: Insulin-like growth factor: A model for tertiary structure accounting for immunoreactivity and receptor binding. Proc Natl Acad Sci USA 1978;75:180–184.
29 Hintz RL: Role of growth-hormone- and insulin-like growth-factor-binding proteins. Horm Res 1990;33:105–110.
30 Drop SLS, Valiquette G, Guyda HJ, et al: Partial purification and characterization of a binding protein for insulin-like activity (ILAs) in human amniotic fluid: A possible inhibitor of insulin-like activity. Acta Endocrinol 1979;90:505–518.
31 Chochinov RH, Mariz IK, Hajek AS, Daughaday WH: Characterization of a protein in mid-term human amniotic fluid which reacts in the somatomedin-C radioreceptor assay. J Clin Endocrinol Metab 1977;44:902–908.
32 Baxter RC, Martin JL: Binding proteins for the insulin-like growth factors: Structure, regulation and function. Prog Growth Factor Res 1989;1:49–68.

33 Hardouin S, Hossenlop P, Segovia B, et al: Heterogeneity of insulin-like growth factor binding proteins and relationships between structure and affinity. 1. Circulating forms in man. Eur J Biochem 1987;170:121–132.
34 Rechler MM, Nissley SP: The nature and regulation of the receptors for insulin-like growth factors. Ann Rev Physiol 1985;47: 425–442.
35 Lobel P, Dahms NM, Breitmeyer J, et al: Cloning of the bovine 215-kDa cation-independent mannose 6-phosphate receptor. Proc Natl Acad Sci USA 1987;84:2233–2237.
36 Cascieri MA, Saperstein R, Hayes NS, et al: Serum half-life and biological activity of mutants of human insulin-like growth factor I which do not bind to serum binding proteins. Endocrinology 1988;123:373–381.
37 Bayne ML, Applebaum J, Underwood D, et al: The C region of human insulin-like growth factor (IGF) I is required for high affinity binding to the type 1 IGF receptor. J Biol Chem 1988; 264:11004–11008.
38 Cascieri MA, Chicchi GG, Applebaum J, et al: Mutants of human insulin-like growth factor I with reduced affinity for the type I insulin-like growth factor receptor. Biochemistry 1988;27: 3229–3233.
39 Jansen J, van Buul-Offers SC, Hoogerbrugge CM, Van den Brande JL: Effects of a single cleavage in insulin-like growth factors I and II on binding to receptors, carrier proteins and antibodies. Biochem J 1990;266:513–520.
40 Cascieri MA, Chicchi GG, Applebaum J, et al: Structural analogs of human insulin-like growth factor (IGF) I with altered affinity for type 2 IGF receptors. J Biol Chem 1989;264:2199–2202.
41 Bayne ML, Applebaum J, Chicchi GG, et al: Structural analogs of human insulin-like growth factor I with reduced affinity for serum binding proteins and the type 2 insulin-like growth factor receptor. J Biol Chem 1988;263:6233–6239.
42 Carlsson-Skwirut C, Lake M, Hartmanis M, et al: A comparison of the biological activity of the recombinant intact and truncated insulin-like growth factor 1 (IGF-I). Biochim Biophys Acta 1989;1011:192–197.
43 Zapf J, Schoenle E, Froesch ER: In vivo effects of the insulin-like growth factors (IGFs) in the hypophysectomized rat: Comparison with human growth hormone and the possible role of the specific IGF carrier proteins. Proc CIBA Found Symp No. 116. London, Pitman Avon. The Bath Press, 1985, pp 169–187.
44 Zapf J, Schoenle E, Jagars E, et al: Inhibition of the action of nonsuppressible insulin-like activity on isolated rat fat cells by binding to its carrier protein. J Clin Invest 1979;63:1077–1084.
45 Busby WH, Klapper DG Jr, Clemmons DR: Purification of a 31,000-dalton insulin-like growth factor binding protein from human amniotic fluid. J Biol Chem 1988;263:14203–14210.
46 De Mellow JSM, Baxter RC: Growth hormone-dependent insulin-like growth factor (IGF) binding protein both inhibits and potentiates IGF-I-stimulated DNA synthesis in human skin fibroblasts. Biochem Biophys Res Commun 1988;156:199–204.
47 Hiraki Y, Inoue H, Kato Y, et al: Combined effects of somatomedin-like growth factors with fibroblast growth factor or epidermal growth factor in DNA synthesis in rabbit chondrocytes. Mol Cell Biochem 1987;76:185–193.
48 Schoenle E, Zapf J, Humbel RE, Froesch ER: Insulin-like growth factor I stimulates growth in hypophysectomized rats. Nature 1982;296:252–253.
49 Van Buul-Offers SC, Ueda I, Van den Brande JL: Biosynthetic somatomedin C (SM-C/IGF-I) increases the length and weight of Snell dwarf mice. Pediatr Res 1986;20:825–827.
50 Guler H-P, Eckardt K-U, Zapf J, et al: Insulin-like growth factor I increases glomerular filtration rate and renal plasma flow in man. Acta Endocrinol 1989;121:101–106.
51 Maes M, Amans Y, Underwood LE: Decreased serum insulin-like growth factor I response to growth hormone in hypophysectomized rats fed a low protein diet: Evidence for a postreceptor defect. Acta Endocrinol 1988;117:320–326.
52 Price DA, Wit JM, van Buul-Offers SC, et al: Serum somatomedin activity and cartilage metabolism in acutely fasted, chronically malnourished and refed rats. Endocrinology 1979;105: 851–861.
53 Trippel SB, Corvol MT, Dumontier MF, et al: Effect of somatomedin-C/insulin-like growth factor I and growth hormone on cultured growth plate and articular chondrocytes. Pediatr Res 1989;25:76–82.
54 McCusker RH, Campion DR, Kelly Jones W, Clemmons DR: The insulin-like growth factor-binding proteins of porcine serum: Endocrine and nutritional regulation. Endocrinology 1989; 125:505–509.
55 Zapf J, Hauri C, Waldvogel M, et al: Recombinant human insulin-like growth factor I induces its own specific carrier protein in hypophysectomized and diabetic rats. Proc Natl Acad Sci USA 1989;86:3813–3817.
56 Suikkari A-M, Koivisto VA, Rutanen E-M, et al: Insulin regulates the serum levels of low molecular weight insulin-like growth factor-binding protein. J Clin Endocrinol Metab 1988; 66:266–272.
57 Brismar K, Gutniak M, Povoa G, et al: Insulin regulates the 35 kDa binding protein in patients with diabetes mellitus. J Endocrinol Invest 1988;11:599–602.
58 Holly JMP, Smith CP, Dunger DB, et al: Levels of the small insulin-like growth factor-binding protein are strongly related to those of insulin in prepubertal and pubertal children but only weakly so after puberty. J Endocrinol 1989;121:383–387.
59 Rabinowitz D, Merimee TJ, Burgess JA: Growth hormone-insulin interaction: Fact and speculation. Diabetes 1966;15:905–910.
60 Davenport ML, Svoboda ME, Koerber KL, et al: Serum concentrations of insulin-like growth factor II are not changed by short term fasting and refeeding. J Clin Endocrinol Metab 1988;67: 1231–1236.
61 Green H, Morikawa M, Nixon T: A dual effector theory of growth hormone action: Models and hypotheses. Differentiation 1985;29:195–198.
62 Isaksson OGP, Lindahl A, Nilsson A, Isgaard J: Mechanism of the stimulatory effect of growth hormone on longitudinal bone growth. Endocr Rev 1987;8:426–438.
63 Wit JM, Fokker MH, De Muinck Keizer-Schrama SMPF, et al: Effects of two years of methionyl growth hormone therapy in two dosage regimens in prepubertal children with short stature, subnormal growth rate, and normal growth hormone response to secretagogues. J Pediatr 1989;115:720–725.

Dr. J.L. Van den Brande
Wilhelmina Children's Hospital
PO Box 18009
NL–3501 CA Utrecht (The Netherlands)

Horm Res 1990;33:69–75

Hormone Epidermal Growth Factor Interactions in Development[1]

D.A. Fisher

UCLA School of Medicine, Harbor-UCLA Medical Center, Torrance, Calif., USA

Key Words. Epidermal growth factor · Transforming growth factor-α · Amphiregulin · Thyroid hormone · Gonadal steroids · Growth · Development

Abstract. Epidermal growth factor (EGF) is the most important member of a family of growth factors which exert their effects via a single 170,000 M_r plasma membrane receptor. Other members include transforming growth factor-α (TGF-α), amphiregulin and several viral growth factors. The receptor is widely distributed in fetal and postnatal tissues. The predominant family member in the fetus appears to be TGF α. EGF production in tissues matures in the perinatal period. Activation of the receptor in the fetal and neonatal periods in rodents evokes important growth and development actions. Tissue EGF and EGF receptor concentrations are modulated by thyroid hormones, estrogen, testosterone and growth hormone, suggesting that selected growth and developmental actions of thyroid and steroid hormones may be mediated by EGF.

Introduction

Epidermal growth factor (EGF) is the ancestral member of a family of growth factors now known to include transforming growth factor-α (TGF-α), amphiregulin and the pox virus growth factors [1–6]. All have in common a similar amino acid sequence, including 6 cysteine residues with a spacing pattern [4]:

$CX_7\ CX_4\ CX_{10}\ CX_1\ CX_8\ C$,

and all have the ability to bind to a common 170,000 relative molecular weight (M_r) glycoprotein receptor referred to as the EGF receptor [1–6]. EGF is a 53-amino-acid, 6,000 M_r polypeptide derived in the mouse from a 1,217-amino-acid precursor and in humans from a 1,207-amino-acid prohormone [7–9]. TGF-α is a 50-amino-acid polypeptide produced from a 160-amino-acid precursor in humans and from a 154-amino-acid prohormone in rats [10, 11]. Amphiregulin is a 9,000–10,000 M_r polypeptide consisting of 78 (in a truncated form) or 84 amino acids [4]. The prohormone has not yet been characterized [4]. Vaccinia virus growth factor is an approximately 77-amino-acid polypeptide secreted as a 23,000 M_r glycoprotein [3]. Both the EGF and TGF-α prohormone molecules have hydrophobic transmembrane domains near the carboxy terminus, and there is evidence that either prohormone can be produced as in integral membrane protein [12, 13].

EGF and other members of the EGF family of growth factors bind to the cell surface EGF receptor stimulating the intrinsic tyrosine kinase activity, and tyrosine phosphorylation is a critical event in signal transduction [14, 15]. EGF receptors have been described in a variety of tissues and EGF is a potent mitogen for ectodermal and mesodermal cell types [1, 16]. The developmental actions of EGF include stimulation of: palatal development in the fetus; eyelid opening and tooth eruption in neonatal rodents; lung maturation in fetal lambs and rabbits; gastrointestinal, liver and pancreatic matura-

[1] Research supported by USPHS grant NHO755 from the National Institute of Child Health and Human Development.

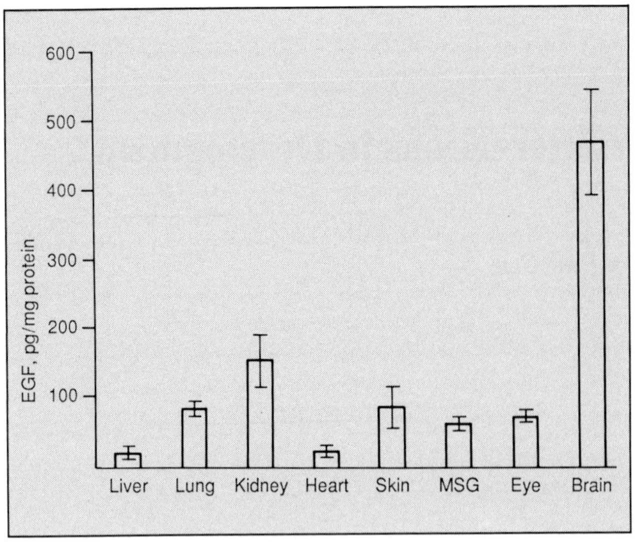

Fig. 1. EGF concentrations in neonatal mouse tissues. Measurements were conducted on day 9 by specific radioimmunoassay of tissue homogenates. Each bar represents the mean ± SEM of 3 or 4 individual determinations. MSG = Moure salivary gland. [Data from ref. 38.]

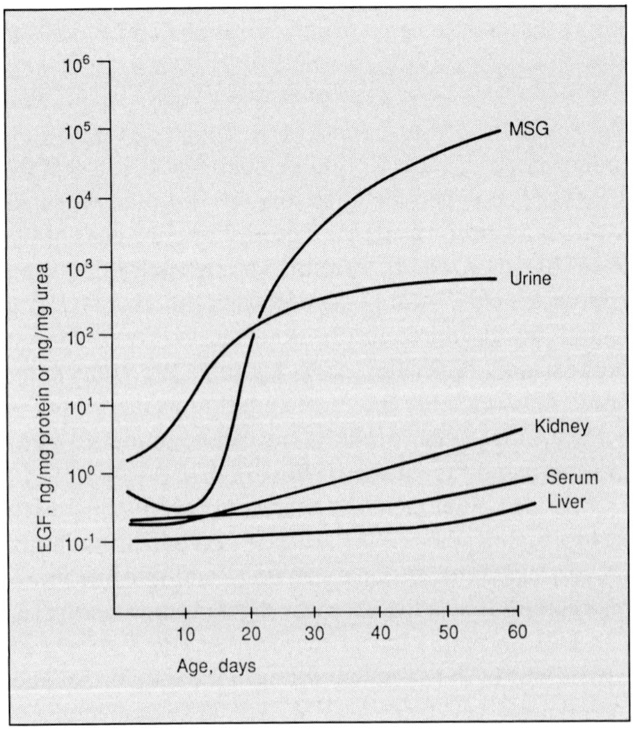

Fig. 2. Ontogenesis of EGF concentrations in mouse salivary gland (MSG), kidney, urine, liver and serum. Values are shown in nanograms per milligram protein on a logarithmic scale. [Data summarized from ref. 37–40.]

tion; thyroid and adrenal gland growth; and pituitary hormone secretion including adrenocorticotropin, growth hormone (GH) and prolactin [1, 16–26]. In adults, EGF appears to have a largely cytoprotective role inhibiting gastric acid secretion, augmenting healing of ulcers, accelerating wound healing and perhaps maintaining integrity of urinary tract epithelial surfaces [21, 27].

EGF and TGF-α in the Fetus

Most of the information regarding the role of EGF in mammalian development has been developed in rodents, especially mice. EGF receptors are present in the fetal mouse although EGF and EGF precursor mRNA levels are absent or low in most fetal rodent tissues [28–31]. TGF-α prohormone mRNA is present early in fetal life, particularly in decidual tissue of the mouse [32, 33]. Moreover, fetal tissues have higher levels of EGF-receptor-binding activity than EGF immunoreactivity [28]. Thus, it is proposed that the ligand for the fetal EGF receptor in mice is TGF-α. Similar data have been developed for the fetal sheep [34]. Recent information indicating the presence of TGF in neonatal rat tissues including brain, lung, liver and kidney suggest that TGF-α may be important in neonatal development [35].

EGF Ontogenesis in the Rodent

Tissue levels of EGF in the neonatal mouse are low, in the 5- to 200-pg/mg protein range (fig. 1). One exception is brain tissue which manifests an EGF concentration approximating 500 pg/mg protein [36]. The maturation patterns of tissue and blood levels of EGF in selected neonatal mice are summarized in figure 2. Levels are low in the early neonatal period and increase progressively in most tissues during the first 30–60 days [36–40]. EGF concentrations in mouse salivary glands increase from levels of 50–70 pg/mg protein during the first 2 weeks to 100 ng/mg protein by 3 weeks. There is a further increase thereafter, more marked in males than females. Adult levels in males at 70 days approximate 30 μg/mg protein; female levels are 50- to 100-fold lower [37]. Urine levels increase from 1–2 ng/mg urea in the neonatal period to 100 ng/mg urea by 3 weeks, and kidney concentrations increase from 100–150 to 300–400 pg/mg protein during the first 3 weeks. Adult values approximate 1 μg/mg urea and 150 ng/mg protein, but in contrast to salivary gland

EGF, urine and kidney EGF concentrations are higher in females than in males [38]. EGF levels in ocular tissue (not shown) increase 10-fold during the first 10 days [36]. EGF concentrations in other tissues and in serum increase more modestly. Schaudies et al. [41] have reported an increase in fasting gastrointestinal EGF levels of 3- to 8-fold during the first 2–4 weeks.

Hormonal Effects on EGF and EGF Receptors

The dramatic increases in ocular tissue, urine and salivary gland EGF concentrations have been shown to correlate with the increase in serum thyroid hormone levels in the mouse during the first 1–3 weeks of postnatal life [37]. Moreover, it has been shown that the delayed eye opening in hypothyroid neonatal rodents can be prevented by administration of exogenous EGF [42], and Gresik and Barka [43] have shown an effect of thyroxine (T_4) on salivary gland EGF levels in the neonatal period. These observations suggested the hypothesis that thyroid hormones modulate EGF production in developing rodents and that EGF mediates at least some of the developmental actions of thyroid hormones.

To test this hypothesis T_4 was administered to newborn mice for varying periods during the first 2 weeks of life. The hormone stimulated precocious eyelid opening and precocious increases in EGF concentrations in ocular tissue, skin, kidney, urine and salivary glands [44–48]. Thyroid hormone also increased EGF receptor binding in skin of normal mice, and congenital hypothyroidism delayed the appearance of EGF receptors in neonatal mouse liver [49, 50]. Thyroid hormone markedly increased EGF receptor and receptor mRNA levels in young adult liver [51]. T_4 had a limited effect on liver EGF concentrations, no effect on liver pro-EGF mRNA levels, and no consistent effect on plasma or serum EGF concentrations [39, 51, 52]. Moreover, there was a significant correlation between the ontogenic increases in liver and plasma EGF [39]. These latter data suggest that circulating EGF is derived largely from liver and perhaps other T_4 independent sources.

Other hormones also have been shown to influence tissue EGF concentrations and EGF receptor binding. Testosterone increases EGF concentrations in salivary gland tissue, GH increases urine EGF levels, and estradiol increases EGF and pro-EGF mRNA concentrations in urine, kidney and uterine tissue of the mouse [30, 43, 53–58]. In adult animals, androgens and GH increase EGF receptor binding in liver and prostate, and estrogen increases EGF receptor and EGF receptor mRNA levels in uterine tissue of the mouse [59–64].

Table 1. Summary of effects of thyroid hormone, androgens, estrogens and GH on tissue EGF concentrations and EGF receptor binding in selected rodent tissue

Tissue	Hormone			
	thyroid	androgen	estrogen	GH
MSG	↑EGF[1]	↑EGF[1]		
Thyroid				
Kidney	↑EGF[1] ?EGFR			
Urine	↑EGF			↑EGF
Liver	↑EGFR[1]	↑EGFR		↑EGFR
Skin	↑EGF ↑EGFR			
Ocular tissue	↑EGF ?EGFR			
Uterus			↑EGF[1] ↑EGFR[1]	
Prostate		↑EGF ↑EGFR		

MSG = Mouse salivary gland; EGFR = EGF receptor.
[1] mRNA levels also measured; parallel changes in pro-EGF and EGF receptor mRNA levels.

These data, summarized in table 1, support the hypothesis that EGF mediates the growth and developmental actions of thyroid hormones early in development and that steroid hormones become important at the time of puberty. There is no information regarding hormonal effects on tissue TGFα concentrations.

Differential Tissue Processing of EGF Prohormone

The EGF precursor molecule in the mouse, as indicated, is composed of 1,217 amino acids; EGF accounts only for 53 of these [7, 8]. Pro-EGF mRNA and EGF immunoreactivity colocalize in salivary gland tissue to the granular tubules, and chromatographic analysis of pro-EGF-processed species in salivary tissue reveals a predominant 6,000 M_r (mature EGF) species, several intermediate-size peptides and only minimal amounts of 130,000 M_r or larger protein [27]. There are 7 other EGF-like domains in the pro-EGF molecule [7, 8], but whether any of these has biological significance is not yet clear. Little is known at present regarding the control of pro-EGF processing.

Pro-EGF can also be produced as a membrane-bound protein in transfected cells [13], and Rall et al. [27] reported that the predominant pro-EGF-derived protein species in the kidney approximates 130,000 M_r, the size predicted for the EGF prohormone from the cDNA structure. Pro-EGF mRNA and EGF-like immunoreactivity colocalize in the kidney to the thick ascending loop of Henle (TALH) and the distal collecting tubule (DCT) [65–67]. By electron microscopy EGF immunoreactivity localizes to the luminal membrane of the TALH and DCT [68]. In the mouse, EGF excretion exceeds urea clearance by 150-fold, suggesting predominant renal secretion [69]. Moreover, salivary gland excision has no effect on urine EGF excretion, and thyroid hormone administration increases urine EGF excretion without altering plasma EGF levels [69]. Testosterone, in contrast, increases salivary gland and plasma EGF levels, but has no effect on urine EGF excretion [65].

In humans urine EGF concentrations, expressed per milligram creatinine, increase progressively during infancy to a peak between 1 and 2 years of age with a plateau thereafter through 20 years [70, 71]. Circulating EGF in humans, in contrast to mice, is largely confined to blood platelets and is released in the process of blood coagulation [72, 73]. Human plasma contains little EGF, and urine EGF in humans, as in mice, is largely accountable on the basis of renal secretion rather than glomerular filtration [73].

Recent data for the mouse mammary gland show cellular colocalization of pro-EGF mRNA and EGF-like immunoreactivity to the secretory ducts and cellular localization of EGF immunoreactivity in trypsin-treated specimens to the luminal border of the secretory alveolar cells [74]. These data support the view that pro-EGF is produced as a membrane protein in mammary gland as in kidney. Thus, in salivary gland, pro-EGF is processed to 6,000 M_r mature EGF and stored for secretion into saliva. Secretion is regulated, at least in part, by plasma membrane receptor-mediated events. In kidney and perhaps mammary gland, there is little or no storage of EGF and data are consistent with the view that pro-EGF is produced as a membrane-bound protein which is then degraded to 6,000 M_r EGF by membrane or luminal proteases. This mechanism may serve as an alternative secretory pathway to provide relatively constant levels of EGF in milk and urine, respectively. At present, we have little information regarding the enzymatic mechanisms for cleavage of the membrane-bound EGF precursor.

Hormonal modulation of tissue EGF and EGF receptor concentrations has been clearly demonstrated. Thyroid hormones exert marked effects on both EGF and/or EGF receptor levels in many tissues and fluids during the period of neonatal development in rodents. Steroid hormones increase EGF levels and/or EGF receptor binding in selected tissues postpubertally. The mechanism(s) of these effects is not yet clear. There are effects on pro-EGF and EGF receptor mRNA levels (table 1), but the extent to which these changes reflect transcription or message stabilization is not known. Moreover, the effects of thyroid hormones on the translation of pro-EGF mRNA and pro-EGF processing remain obscure.

References

1 Carpenter G, Cohen S: Epidermal growth factor. Annu Rev Biochem 1979;48:193–216.
2 Hunkapillar MW, Hood LE, Todaro GJ: Rat transforming growth factor type 1: Structure and relation to epidermal growth factor. Science 1984;233:1079–1081.
3 Stroobant P, Rice AP, Gullick WJ, et al: Purification and characterization of vaccinia virus growth factor. Cell 1985;42:383–393.
4 Shoyab M, Plowman GD, McDonald VL, et al: Structure and function of human amphiregulin: A member of the epidermal growth factor family. Science 1989;243:1074–1076.
5 Massague J: Epidermal growth factor-like transforming growth factor. II. Interaction with epidermal growth factor receptor in human placenta and membranes and A431 cells. J Biol Chem 1983;258:13614–13620.
6 Pike LJ, Marquardt H, Todaro GJ, et al: Transforming growth factor and epidermal growth factor stimulate the phosphorylation of a synthetic, tyrosine-containing peptide in a similar manner. J Biol Chem 1982;257:14628–14631.
7 Gray A, Dull TJ, Ullrich A: Nucleotide sequence of epidermal growth factor cDNA predicts a 128,000 molecular weight precursor. Nature (Lond) 1983;303:703–705.
8 Doolittle RR, Feng DF, Johnson MS: Computer based characterization of epidermal growth factor precursor. Nature (Lond) 1984;307:558–560.
9 Bell GI, Fong NM, Stempien MM, et al: Human epidermal growth factor precursor: cDNA sequence, expression in vitro and gene organization. Nucl Acids Res 1986;14:8427–8446.
10 Derynck R, Roberts AB, Winkler ME, et al: Human transforming growth factor-α: Precursor structure and expression in *E. coli*. Cell 1984;38:287–297.
11 Lee DC, Rose TM, Webb NR, Todaro GJ: Cloning and sequence analysis of a cDNA for rat transforming growth factor α. Nature 1985;313:489–491.
12 Wong ST, Winchell LF, McCune BK, et al: The TGFα precursor expressed on the cell surface binds to the EGF receptor on adjacent cells, leading to signal transduction. Cell 1989;56:495–506.
13 Mroczkowski B, Reich M, Chen K, et al: Recombinant human epidermal growth factor is a glycosylated membrane protein with biological activity. Mol Cell Biol 1989;9:2771–2778.
14 Schlessinger J: Allosteric regulation of the epidermal growth factor receptor kinase. J Cell Biol 1987;103:2067–2072.

15 Schlessinger J, Ullrich A, Honegger AM, Moolenaar WH: Signal transduction by epidermal growth factor receptor. Cold Spring Harbor Symp Quant Biol 1988;53:515–519.
16 Gospodarowicz D: Epidermal and nerve growth factors in mammalian development. Annu Rev Physiol 1981;43:251–263.
17 Catterton W, Escobedo MB, Saxon WR, et al: Effect of EGF on lung maturation in fetal rabbit. Pediatr Res 1979;13:104–108.
18 Sundell H, Gray H, Serenius F, et al: Effects of epidermal growth factor on lung maturation in fetal lambs. Am J Pathol 1980;100:707–726.
19 Gross I, Dynia D, Rooney S, et al: Influence of epidermal growth factor on fetal rat lung development in vitro. Pediatr Res 1986;20:473–477.
20 Partanen AM. Epidermal growth factor and tooth development. Proc Finn Dent Soc 1987;83:181–190.
21 Marti V, Burwen SJ, Jones AL. Biological effects of epidermal growth factor with emphasis on the gastrointestinal tract and liver: An update. Hepatology 1989;9:126–138.
22 Chatterjee A, Sieradzki J, Schatz H: Epidermal growth factor stimulates (pro)insulin biosynthesis and ^{3}H-thymidine incorporation in isolated pancreatic rat islets. Horm Metab Res 1986;18:873–874.
23 Polk DH, Ervin MG, Padbury JF, et al: Epidermal growth factor acts as a corticotropin releasing factor in chronically catheterized fetal lambs. J Clin Invest 1987;79:984–988.
24 Murdoch GH, Porter E, Nicolaisen AK, et al: Epidermal growth factor rapidly stimulates prolactin gene transcription. Nature (Lond) 1982;300:192–194.
25 Luger A, Calogero A, Kalogeras K, et al: Interaction of epidermal growth factor with the hypothalamic-pituitary-adrenal axis: Potential physiologic relevance. J Clin Endocrinol Metab 1988;66:334–337.
26 Heldin NE, Westermark B: Epidermal growth factor, but not thyrotropin, stimulates the expression of c-fos and c-myc messenger ribonucleic acid in porcine thyroid follicle cells in primary culture. Endocrinology 1988;122:1042–1046.
27 Rall LB, Scott J, Bell BI, et al: Mouse pre pro-epidermal growth factor synthesis by the kidney and other tissues. Nature (Lond) 1985;313:228–231.
28 Nexo E, Hollenberg M, Figueroa A, Pratt RM: Detection of EGF-urogastrone and its receptor in fetal mouse development. Proc Natl Acad Sci USA 1980;77:2782–2785.
29 Adamson ED, Meek J: Epidermal growth factor receptors during mouse development. Dev Biol 1984;103:62–70.
30 Gubits RM, Shaw PA, Gresick EW, et al: Epidermal growth factor gene expression is regulated differently in mouse kidney and submandibular gland. Endocrinology 1986;119:1382–1387.
31 Popliker M, Shatz A, Avini A, et al: Onset of endogenous synthesis of epidermal growth factor in neonatal mice. Dev Biol 1987;119:38–44.
32 Lee DC, Rochford R, Todaro GJ, et al: Developmental expression of rat transforming growth factor alpha mRNA. Mol Cell Biol 1985;5:3644–3646.
33 Han VKM, Hunter FS III, Pratt RM, et al: Expression of rat transforming growth factor alpha mRNA during development occurs predominantly in the maternal decidua. Mol Cell Biol 1987;7:2335–2343.
34 Freemark M, Comer M: Epidermal growth factor (EGF)-like transforming growth factor (TGF)-activity and EGF receptors in ovine fetal tissues: Possible role for TGF in ovine fetal development. Pediatr Res 1987;22:609–615.
35 Brown PI, Lam, R, Lakshmanan J, Fisher DA: Transforming growth factor alpha (TGF-α) in the developing rat. Am J Physiol, in press.
36 Lakshmanan J, Perheentupa J, Hoath SB, et al. Epidermal growth factor in mouse ocular tissue: Effects of T4 and exogenous EGF. Pediatr Res 1985;19:315–319.
37 Walker P, Weichsel ME Jr, Eveleth D, Fisher DA: Ontogenesis of NGF and EGF in submaxillary gland and NGF in brain of immature male mice: Correlations with ontogenesis of serum levels of thyroid hormones. Pediatr Res 1982;16:520–524.
38 Perheentupa J, Lakshmanan J, Fisher DA: Urine and kidney EGF: Ontogeny and sex differences in the mouse. Pediatr Res 1985;19:428–432.
39 Laborde NP, Grodin M, Buenaflor G, et al: Ontogenesis of epidermal growth factor in liver of BALB mice. Am J Physiol 1988;18:E28–E32.
40 Perheentupa J, Lakshmanan J, Hoath SB, et al: Epidermal growth factor measurements in mouse plasma: Method, ontogeny and sex difference. Am J Physiol 1985;11:E391–E396.
41 Schaudies RP, Grimes J, Davis D, et al: EGF content in the gastrointestinal tract of rats: Effect of age and fasting/feeding. Am J Physiol 1989;19:G856–G861.
42 Hamburgh M, Mendoza LA, Burkart JF, Weil F: Thyroid dependent processes in the developing nervous system; in Hamburgh M, Barrington EJW (eds): Hormones in Development. New York, Appleton Century Crofts, 1971, pp 403–415.
43 Gresik EW, Barka T: Precocious development of granular convoluted tubules in the mouse submandibular gland induced by thyroxine or by thyroxine and testosterone. Am J Anat 1980;159:177–185.
44 Hoath SB, Lakshmanan J, Scott SM, Fisher DA: Effect of thyroid hormone on EGF concentration in neonatal mouse skin. Endocrinology 1983;112:308–314.
45 Hoath S, Lakshmanan J, Fisher DA: Thyroid hormone effects on skin and hepatic EGF concentrations in neonatal mice. Biol Neonate 1984;45:49–52.
46 Scott S, Alm J, Fisher DA: Effect of thyroid hormone and sex status on EGF concentrations in the submandibular gland of congenitally hypothyroid mouse model. J Dev Physiol 1986;8:461–466.
47 Lakshmanan J, Perheentupa J, Macaso T, Fisher DA: Acquisition of urine, kidney and submandibular gland epidermal growth factor responsiveness to thyroxine administration to neonatal mice. Acta Endocrinol 1985;109:511–516.
48 Perheentupa J, Lakshmanan J, Fisher DA: Epidermal growth factor in neonatal mouse urine: Maturative effect of thyroxine. Pediatr Res 1984;18:1080–1084.
49 Hoath SB, Lakshmanan J, Fisher DA: Epidermal growth factor binding to neonatal mouse skin explants and membrane preparations: Effect of triiodothyronine. Pediatr Res 1985;19:277–280.
50 Alm J, Scott S, Fisher DA: Epidermal growth factor receptor ontogeny in mice with congenital hypothyroidism. J Dev Physiol 1986;8:377–385.
51 Laborde NP, Buenaflor G, Grodin M, Fisher DA: Thyroid hormone effects on liver epidermal growth factor metabolism in the mouse (abstract 471). Pediatr Res 1988;23:279A.
52 Perheentupa J, Lakshmanan J, Hoath SB, et al: Hormonal modulation of mouse plasma concentrations of epidermal growth factor. Acta Endocrinol 1984;107:571–576.

53 Walker P, Weichsel ME Jr, Hoath SB, et al: Effect of T4, testosterone and corticosterone on NGF and EGF concentrations in adult female mouse SMG: Dissociation of EGF and NGF responses. Endocrinology 1981;109:582–587.
54 Perheentupa J, Lakshmanan J, Macaso T, Fisher DA: Growth hormone increases neonatal mouse urine epidermal growth factor. Acta Endocrinol 1984;106:184–189.
55 Gonzales F, Lakshmanan J, Hoath S, Fisher DA: Effect of 17 beta-estradiol on uterine epidermal growth factor concentration in immature mice. Acta Endocrinol 1984;105:425–428.
56 Korach KS, McLachlan JA, Teng CT: Influence of estrogens in mouse uterine epidermal growth factor precursor protein and messenger ribonucleic acid. Endocrinology 1988;122:2355–2363.
57 Tuomela T, Viinikka L, Perheentupa J: Mouse epidermal growth factor concentrations are altered by gonadectomy and treatments with estradiol and progesterone. Life Sci 1989;44:1815–1821.
58 DiAugustine RP, Petrusz P, Bell GI, et al: Influence of estrogens on mouse uterine epidermal growth factor precursor protein and messenger ribonucleic acid. Endocrinology 1988;122:2355–2363.
59 Jansson JO, Ekberg S, Hoath S, et al: Growth hormone enhances hepatic epidermal growth factor receptor concentration in mice. J Clin Invest 1988;82:1871–1876.
60 Alm J, Lakshmanan J, Fisher DA: Sexual dimorphism and testosterone effects on liver epidermal growth factor receptors in mice. Acta Endocrinol 1986;113:140–144.
61 Schuurmans ALG, Bolt J, Mulder E: Androgens and transforming growth factor-β modulate the growth response to epidermal growth factor in human prostatic tumor cells. Mol Cell Endocrinol 1988;60:101–104.
62 Abdulmaged MT, Notiz HH: Prostatic epidermal growth fator receptors and their regulation by androgens. Endocrinology 1987;121:1461–1467.
63 Gardner RM, Verner G, Kirkland JL, Stancel GM: Regulation of uterine epidermal growth factor (EGF) receptors by estrogen in the mature rat and during the estrus cycle. J Steroid Biochem 1989;32:339–343.
64 Lingham RB, Stancel GM, Loose Mitchell DS: Estrogen regulation of epidermal growth factor receptor messenger ribonucleic acid. Mol Endocrinol 1988;2:230–235.
65 Fisher DA, Salido EG, Barajas L: Epidermal growth factor and the kidney. Annu Rev Physiol 1989;51:67–80.
66 Salido EC, Fisher DA, Barajas L: In situ hybridization of epidermal growth factor mRNA in mouse kidney. Am J Physiol 1989; 19:F632–F638.
67 Salido EC, Barajas L, Lechago J, et al: Immunocytochemical localization of epidermal growth factor in mouse kidney. J Histochem Cytochem 1986;34:1155–1160.
68 Salido EC, Fisher DA; Barajas L: Immunoelectron microscopy of epidermal growth factor in mouse kidney. J Ultrastruct Mol Struct Res 1986;91:105–113.
69 Perheentupa J, Lakshmanan J, Fisher A: Epidermal growth factor in mouse urine: Non-blood origin and increase by sialoadenectomy and T4 therapy. Acta Endocrinol 1985;108:428–432.
70 Matilla AL, Perheentupa J, Personen K, Vinikka L: Epidermal growth factor in human urine from birth to puberty. J Clin Endocrinol Metab 1975;61:997–1000.
71 Matilla AL: Human urinary epidermal growth factor: Effects of age, sex and female endocrine status. Life Sci 1986;39:1879–1884.
72 Oka Y, Orth DN: Human plasma epidermal growth factor urogastrone is associated with blood platelets. J Clin Invest 1983; 72:249–259.
73 Callegari C, Laborde NP, Buenaflor G, et al: The source of urinary epidermal growth factor in humans. J Appl Physiol 1988; 58:26–31.
74 Brown CF, Teng CT, Pentecost BT, Di Augustine RP: Epidermal growth factor precursor in mouse lactating mammary gland alveolar cells. Mol Endocrinol 1989;3:1077–1083.

Prof. D.A. Fisher
UCLA School of Medicine
Harbor-UCLA Medical Center
1000 West Carson Street
Torrance, CA 90509 (USA)

Discussion

M.T. Corvol, France: When you showed that T_4 was increasing the EGF content, EGF mRNA and EGF receptor maturation in several tissues, did you have any idea about the level of action of T_4: is it on pro-EGF processing or at the transcriptional level?

D.A. Fisher, USA: The mechanism of the T_4 effect on pro-EGF or EGF receptor mRNA levels is not yet clear. There may be effects on transcription or on mRNA half-life.

M.T. Corvol: You showed two different ways of EGF action, one corresponding to direct secretion of pro-EGF and another on producing mature peptide. Do you think that T_4 may induce the secretion of the mature EGF?

D.A. Fisher: T_4 increases pro-EGF production via the regulated or secretory pathway in the salivary gland and seems to increase pro-EGF and EGF via the constitutive or membrane pathway in the kidney. I would assume that the cell of origin determines the route of production. However, in a secretory-type cell, constitutive processing of pro-EGF could occur. We have no examples of such dual production to data but in such a cell T_4 might induce one pathway versus another.

A.W. Root, USA: Are there data on the role of EGF in patients with overgrowth syndromes such as patients with neurofibromatosis or in patients who form keloid scans?

D.A. Fisher: Increased EGF receptors have been shown in active psoriatic lesions as well as acanthosis nigricans, seborrheic keratoses and skin tags. Increased EGF receptor levels have also been associated with verucca vulgaris and molluscum contagiosum as well as with squamous cell carcinomas from several locations. TGF-α expression is also increased in psoriatic lesions.

A. Cotterill, UK: Does the low-density lipoprotein (LDL) receptor have any role in the release of EGF from the cell surface expressed prepro-EGF? Does LDL binding enhance release?

D.A. Fisher: No. The LDL receptor has no known role in EGF metabolism. The sequence homology between pro-EGF and the LDL receptor was very intriguing, and with the transmembrane domain suggested a possible membrane receptor function for pro-EGF in the kidney. The evolutionary significance of the homology is not clear.

J. Nerup, Denmark: You described two possible mechanisms for EGF/EGF receptor action: (a) EGF receptor tyrosine kinase use lipocortin I as a substrate, and (b) the intriguing observation of Hillary Kaprowski that EGF might be internalized and transported to the nucleus to act as a DNA-binding protein. Do you see these mechanisms as mutually exclusive or as complementary? Would it not make sense if the mitotic signal was separated from the other effects of EGF?

D.A. Fisher: The significance of a possible nuclear DNA action or binding of EGF is not clear. Certainly the two mechanisms – phosphorylative activation or inhibition of intracellular messenger molecules and direct nuclear EGF receptor binding – are not mutually exclusive. And one or more substrates for tyrosine phosphorylation may be involved in selected cell types. These include lipocortins, phospholipase C11, PP42 and perhaps others. It is believed that tyrosine kinase activation is essential for EGF action. If this is the case the mechanisms would have to be complementary.

Horm Res 1990;33:76–86

0301–0163/90/0334–0076$2.75/0

The Gonadotropin-Releasing Hormone Receptor: Signals Involved in Gonadotropin Secretion and Biosynthesis

H. Dan-Cohen, D. Ben-Menahem, Z. Naor

Department of Biochemistry, George S. Wise Faculty of Life Sciences, Tel Aviv University, Tel Aviv, Israel

Key Words. Gonadotropin · Protein kinase C · Phosphoinositide turnover · G protein · LH mRNA · Gene expression

Abstract. The neurohormone gonadotropin-releasing hormone (GnRH) is a decapeptide which is synthesized in the hypothalamus and released into the hypophysial portal system in a pulsatile manner. GnRH exerts its effect on the anterior pituitary gonadotrophs where it regulates the secretion and synthesis of gonadotropins (luteinizing hormone and follicle-stimulating hormone) through receptor-mediated actions. The GnRH receptor has been characterized and shown to be coupled to the formation of 'second messengers' which participate in signal transduction mechanisms. GnRH stimulation of luteinizing hormone release is a Ca^{2+}-dependent process. G protein, phosphoinositide hydrolysis, protein kinase C as well as arachidonic acid and some of its metabolites were identified as possible mediators in the process.

GnRH-Receptor Interaction

When stable analogues of gonadotropin-releasing hormone (GnRH) became available it was possible to set up a reliable radioreceptor assay [1, 2]. Binding studies were then carried out in pituitary membrane preparations (at 4 °C) or cultured pituitary cells (at room temperature and 37 °C), demonstrating high-affinity binding sites for GnRH [2, 3].

Occupancy of about 20% of GnRH-binding sites was sufficient to evoke 80% of the luteinizing hormone (LH) response, pointing out that most of the receptors might be 'spare receptors' [3]. To confirm that the binding sites for GnRH were indeed receptors, binding studies were carried out at physiological conditions (isotonic buffers, 37 °C), and binding affinities were shown to correlate to the biological potencies of various GnRH analogues [4]. GnRH regulates the number of its own receptors (up- and down-regulation) in vivo and in vitro [2, 5–8]. Indeed, changes in receptor number are observed during the reproductive cycle, pregnancy and lactation, and after castration, hypothalamic lesions or administration of GnRH antagonists [2, 9, 10]. The changes in GnRH receptor number are explained by changes in endogenous GnRH level. Similar conclusions were derived from in vitro studies where GnRH was shown to up-regulate its own receptors at low concentrations and to down-regulate the receptors at high concentrations [6, 7].

It is possible that protein kinase C (PKC) is involved in up-regulating GnRH receptors, as addition of the tumor promoter 12-O-tetradecanoylphorbol-13-acetate (TPA) to primary cultures of rat pituitary cells resulted in increased number of GnRH-binding sites [11, 12]. Since GnRH was shown to activate PKC by redistribution of the cytosolic enzyme to the membrane [13, 14], it is possible that PKC is involved in homologous regulation of GnRH receptors [11, 12]. Down-regulation of GnRH receptors represents most likely only one aspect of the phenomenon of desensitization which is caused by prolonged exposure to GnRH in vivo and in vitro. Uncoupling of the GnRH transducer (G protein) and/or the effector (phospholipase C; PLC) is probably the major biochemical process contributing to desensitization.

A 60-kilodalton membrane protein is presumed to be the GnRH receptor as revealed by photoaffinity labeling and ligand immunoblotting [15–17]. Similar binding sites are present in rat gonads and human placenta, but the extrapituitary effects of GnRH and its analogues are still not clear [18–22]. The GnRH receptor was solubilized recently and purified from pituitary membranes, and monoclonal antibodies against the receptor were raised [23–25]. Further studies will, therefore, concentrate on isolation of the GnRH receptor gene. Unlike thyrotropin-releasing hormone or somatostatin which interact with various pituitary cells, GnRH binds to and activates only the gonadotrophs as evident from biochemical, morphological and cytochemical studies [26–29].

The fate of the GnRH-receptor complex was investigated by various methods. Using video-intensified fluorescent microscopy and rhodamine-labeled *D*-lys^6-GnRH analogue it was demonstrated in living cells that initially GnRH is distributed evenly on the gonadotrophs, forms aggregates and finally the complex is internalized [30, 31]. Internalization is not involved in the biological activity of the decapeptide but serves most likely as means of ligang degradation and receptor recycling. In another approach, a *D*-lys^6-GnRH-biotin probe was prepared and its fate was investigated using the avidin-biotin-peroxidase method [29, 32].

To follow the interaction of the probe with pituitary gonadotrophs the cells were separated from other cell types by centrifugal elutriation [28, 32]. Following ligand binding, cellular processes appeared which stained for both the biotin-GnRH probe and also the gonadotropins. It was, therefore, suggested that the endocytotic and exocytotic sites were linked following receptor activation [26]. Formation of the processes may have occurred by the addition of the secretory granule membranes which were reported to contain GnRH receptors [33, 34]. Thus, exocytosis of the granules may have added more receptors for further binding.

It was noted that GnRH binds to the multhihormonal gonadotrophs [LH + follicle-stimulating (FSH) cells $\approx$60% of total gonadotrophs] and also to the monohormonal cells ($\approx$18% LH cells and 22% FSH cells) [26, 35]. Whether preferential release of LH or FSH results from selective interaction of GnRH with monohormonal gonadotrophs should be investigated. Alternatively, gonadal steroids and inhibin might dictate the target cells for GnRH by selective blockade of the multihormonal cells and one group of the monohormonal gonadotrophs.

The combined studies have indicated that GnRH interacts only with the gonadotrophs. The hormone-receptor complex forms patches and internalization occurs in small vesicles. Gonadotropin stores are mobilized to the site of endocytosis for unknown reasons. The endocytotic vesicles may transfer GnRH to the lysosome for degradation, while multivesicular bodies and possibly secretory granules may be responsible for GnRH receptor recycling.

Role of Phosphoinositide Turnover in GnRH-Induced Gonadotropin Release

The receptor-mediated stimulation of gonadotropin release in rat anterior pituitary cells by GnRH is known to be accompanied by increased phospholipid turnover [36–38]. A PLC activates the hydrolysis of phosphatidylinositol-4,5-bisphosphate (PIP_2) to generate 1,2-diacylglycerol (DG) and inositol-1,4,5-trisphosphate (IP_3; fig. 1). Addition of GnRH to pituitary cells prelabeled with myo-[2-^{3}H]-inositol stimulates the appearance of labeled IP_3 in the presence of Li^+ in the medium [37]. Activation of PLC by GnRH does not require elevation of cytosolic free Ca^{2+}. This finding, together with the early appearance of IP_3 ($\sim$10 s), indicates that activation of phosphoinositide turnover is an early event in GnRH-induced LH release. The products of the cycle: DG, IP_3, arachidonate and inositol-1,3,4,5-tetrakisphosphate [IP_4; the product of an $I(1,4,5)P_3$-3-phosphokinase] are regarded as 'second messengers'. DG activates the Ca^{2+}/phospholipid-dependent PKC [39, 40]. IP_3 mobilizes intracellular Ca^{2+} in many systems [41] including pituitary cells [42], and IP_4 has been implicated recently in receptor-mediated Ca^{2+} entry [43–45]. Although IP_4 as well as IP_5 and IP_6 were shown to be elevated by GnRH in pituitary cells [38], their role in GnRH action is not known. IP_3 is converted to inositol, while DG is phosphorylated to phosphatidic acid. Phosphatidic acid can also act as a Ca^{2+} ionophore and is capable of releasing LH by itself. In this way it may participate in the stimulation of gonadotrophs' responses to GnRH [46]. An activated DG and inositol will produce PI, and phosphorylation of PI will result in the formation of PIP and further to PIP_2 (fig. 1).

Involvement of a G Protein in GnRH-Receptor-Activated Phosphoinositide Turnover

It has been proposed that G protein couples cell surface receptors to PLC, in an analogous way to its coupling to the adenylate cyclase system [47]. It was also

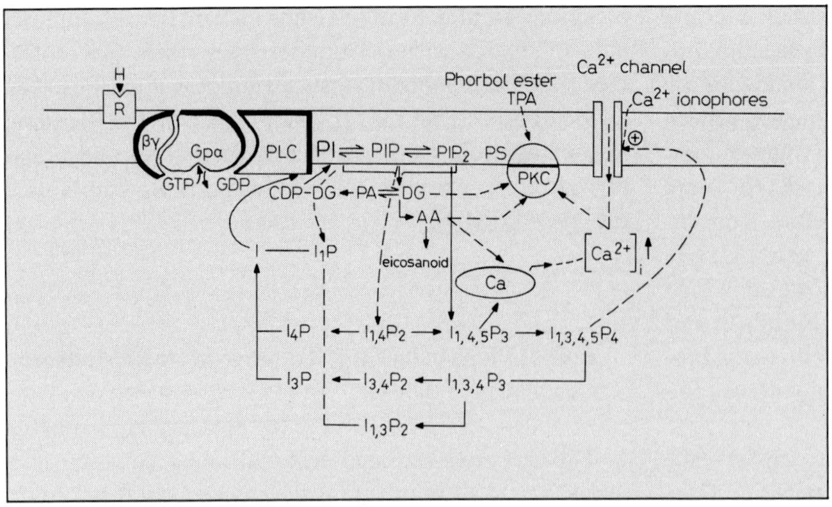

Fig. 1. Phosphoinositide turnover. Hormone (H)-receptor (R) interaction results in dissociation of Gp and activation of PLC. PLC hydrolyzes PIP_2 to form DG and IP_3. DG is then converted to phosphatidic acid (PA) and further to CDP-DG which interacts with free inositol (I) to form PI. PI is phosphorylated to PIP and back to PIP_2. The active isomer of IP_3 mobilizes cellular Ca^{2+} pools, it is then phosphorylated to IP_4 which is thought to be involved in Ca^{2+} influx. IP_3 and IP_4 are converted by successive dephosphorylations to I. DG is also a substrate for DG and monoacylglycerol lipase whose activation will result in the formation of AA and its metabolites, the eicosanoids.

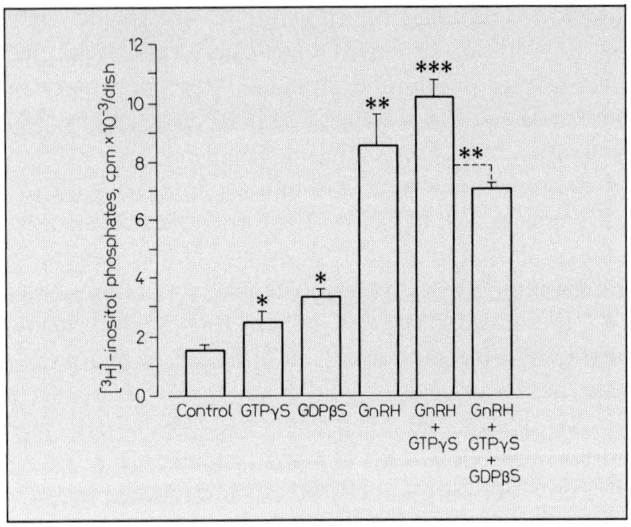

Fig. 2. Effect of GTPγS, GDPβS and GnRH on [^{3}H]-inositol phosphate formation. Pituitary cell cultures were prelabeled with [^{3}H]-inositol. GTPγS (100 μM) and GDPβS (150 μM) were added to ATP^{4-}-permeabilized cells. The resealed cells were treated with or without GnRH (100 nM). * $p < 0.05$; ** $p < 0.01$; *** $p < 0.001$. [From ref. 50.]

suggested that a pertussis-toxin-insensitive G protein termed Gp is involved in inositol phospholipid hydrolysis in pituitary and other cells [37, 48, 49]. We found that addition of GTPγ-S or GnRH to cultured pituitary cells permeabilized by ATP^{4-} resulted in enhanced PLC activity as determined by [^{3}H]-inositol phosphate production [50]. The two effects were not additive, but the combined effect was partially inhibited by GDPβS (fig. 2). GTPγS proved capable of releasing LH in the same cells in a Ca^{2+}-independent manner. In addition, GDPβS, which maintains G proteins in an inactive state, caused a dose-dependent increase in [^{3}H]-inositol phosphate accumulation, which suggests the existence of an inhibitory G protein (Gp_i) that could be involved in inactivation of PLC [50]. A hallmark of G-protein-coupled receptors is the ability of GTP to transfer receptors from a high-affinity to a lower-affinity state. Indeed, GTPγS was found to decrease the binding affinity of GnRH agonist in rat pituitary membrane preparations [50].

Role of Calcium in GnRH Action

GnRH requires Ca^{2+} for its action on gonadotropin secretion [51]. The action of GnRH on cytosolic free Ca^{2+} ($[Ca^{2+}]_i$) was investigated in rat pituitary gonadotrophs preloaded with the fluorescent probe quin-2. GnRH induced an initial rapid increase followed by a prolonged decay of elevated $[Ca^{2+}]_i$ which lasted for several minutes [42]. In single rat anterior pituitary gonadotrophs identified by a reverse hemolytic plaque assay, GnRH induced an initial increase and decrease followed by a secondary prolonged rise and fall of $[Ca^{2+}]_i$ [52]. It seems that internal pools of Ca^{2+} are responsible for the first transient rise in $[Ca^{2+}]_i$ which was reached few seconds after stimulation with GnRH, both in the presence and in the absence of external Ca^{2+} [42, 53, 54]. This cellular Ca^{2+} pool is most probably mobilized via IP_3 formation, since addition of IP_3 to permeabilized pituitary cells resulted in a transient increase in Ca^{2+} which was released from a nonmitochondrial store [53]. The second phase of LH release is mediated in part by influx of Ca^{2+}. The Ca^{2+} influx occurs both via voltage-sensitive (VSCCs) and insensitive Ca^{2+} channels. VSCCs comprise about 50% of the channels. Activation of VSCCs by the dihydrophyridine Ca^{2+} agonist (BAY K8644) stimulated about 35% of the LH release reached by 10 n*M* GnRH. Both Ca^{2+} elevation and GnRH-induced LH release were inhibited similarly (40–50%) by the dihydropyridine Ca^{2+} antagonist, nifedipine. These observations indicate that the second Ca^{2+} pool mobilized by GnRH is derived from extracellular sources via voltage-sensitive and -insensitive Ca^{2+} channels [53, 54].

There is, however, a debate concerning the participation of voltage-insensitive Ca^{2+} channels in GnRH-induced LH release. Whereas Conn et al. [51, 55] proposed that this mobilized Ca^{2+} pool is sufficient to mediate the full response to GnRH, we and others found only a partial role for this pool [52, 53, 56]. The mechanism by which the VSCCs are activated is unclear, since electrophysiological studies have shown that GnRH does not depolarize gonadotrophs [57]. The activation of these Ca^{2+} channels may require therefore the intervention of a diffusible second messenger, the identity of which is not yet known.

Since ionomycin can mobilize cellular Ca^{2+} (like IP_3) and also cause Ca^{2+} influx, it is possible to use the Ca^{2+} ionophore for pharmacological reconstitution experiments. Indeed, addition of ionomycin in a dose which mimics the rise in $[Ca^{2+}]_i$ induced by GnRH resulted only in partial (45%) reconstitution. The results strongly suggest that Ca^{2+} (from cellular and extracellular sources) is not sufficient to mediate GnRH action on gonadotropin secretion [53].

Shangold et al. [52] reported recently that in single gonadotrophs, identified by reverse hemolytic plaque assay, the $[Ca^{2+}]_i$ responses to GnRH stimulation frequently consisted of several rapid oscillations. These kinds of spontaneous oscillations, that were first observed by Woods et al. [58], exist in different cell types and are often induced by agonists that hydrolyze inositol lipids [59]. Such oscillations are generated in Xenopus oocytes after injection with IP_3. It has been speculated that the oscillations might be due to feedback control of the $[Ca^{2+}]_i$ level or, alternatively, that the cells might employ these oscillations as a way of encoding information. But neither the importance nor the mechanism of generating these oscillations in gonadotrophs or other cells is currently understood [59].

Role of Protein Kinase C

PKC, a Ca^{2+}-activated and phospholipid-dependent protein kinase, which was initially identified as a proteolytically activated enzyme by Nishizuka [39], is thought to play a major role in the control of a wide variety of cellular processes. The importance of this enzyme in signal transduction was fully recognized after the discovery that DG, generated during receptor-mediated hydrolysis of phosphoinositides, is a physiological activator of the enzyme. PKC is thought to be the main receptor for the tumor promoter phorbol esters, and therefore the enzyme is also implicated in transformation [39, 40]. PKC is present in the pituitary [13, 14, 60, 61], where it is mostly soluble (70%) and partly particulate (30%). Stimulation of the pituitary by GnRH results in translocation of PKC from the cytosol to the membrane [13, 14]. Such a redistribution exposes the inactive enzyme to phosphatidylserine (PS) and DG. DG activates the enzyme in the presence of near-physiological concentrations of Ca^{2+} and PS. Similarly, the phorbol ester TPA activates PKC by translocating the cytosolic form of the enzyme to the membrane [62, 63]. Studies with cultured pituitary cells demonstrated that TPA also stimulated LH secretion in a time- and dose-dependent manner and that the simultaneous presence of the Ca^{2+} ionophore A23187 and TPA produced a synergistic response that mimicked the exocytotic response of GnRH [64].

PKC is now recognized as a family of at least 7 closely related subspecies [65, 66]. Recently, type I, II and III PKCs have been identified as products of γ-, β- and α-genes respectively. Type II is a mixture of 2 subspecies determined by βI and βII cDNA sequences and results from alternative splicing of a common mRNA transcript [67, 68]. In order to assess the task of the different PKC subspecies in LH secretion, we added purified activated PKCs to TPA-pretreated (PKC-depleted), digitonin-permeabilized cells. Only β- and α-but not γ-type PKC stimulated LH release (fig. 3) [69]. Since the pituitary most likely expresses only the βII- and α-subspecies [70], it is possible that the exocytotic response observed after GnRH challenge are elicited by the α- and βII- enzymes. Whether PKC is an essential link in the signal transduction cascade of GnRH action is still controversial.

The relatively selective PKC inhibitor, staurosporine, inhibited GnRH-induced LH release with half-maximal inhibition (IC_{50}) of about 80 n*M*. Staurosporine inhibited also TPA-induced LH release with a similar IC_{50}, an observation which further supports a role for PKC in GnRH action [71]. However, studies with PKC-depleted

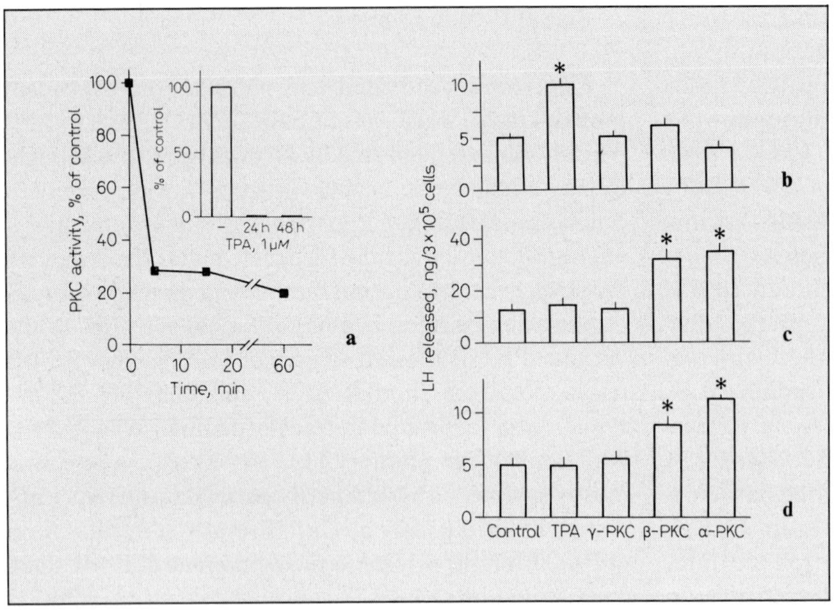

Fig. 3. Insertion of brain PKC subspecies to permeabilized PKC-depleted pituitary cells. **a** Loss of PKC activity in digitonin (16 μ*M*)-permeabilized or TPA-pretreated cells (inset). **b–d** Insertion of brain PKC subspecies (60 ng enzyme) to intact cells (**b**), to digitonin-permeabilized cells (**c**) or TPA-pretreated, digitonin-permeabilized cells (**d**). * $p < 0.01$ [from ref. 69].

pituitary cells in primary tissue culture (exposed for long incubation periods to high concentrations of TPA), showed inconclusive results [71–73]. Some discrepancies between the different responses may be due to the recently reported differential translocation and subsequent 'down-regulation' of PKC subspecies by TPA [74–77]. Therefore, this suggests caution in interpreting PKC 'down-regulation' data. With such caution applied, the findings reported here and elsewhere [71] tend to support the view that PKC activation is a necessary step in the stimulation of LH release by GnRH.

Role of Arachidonic Acid

Arachidonic acid (AA) and its metabolites have also been implicated in GnRH action. GnRH was shown to release AA. Products of 5-lipoxygenase and epoxygenase can release LH: mainly 5-hydroxy-6,8,11,14-eicosatetraenoic acid, the leukotrienes C_4 and B_4, and 5,6-epoxyicosatrienoic acid [78–80]. Lipoxygenase inhibitors such as nordihydroguaiaretic acid and 5,8,11,14-icosatetraynoic acid caused inhibition of GnRH-induced LH release [78]. Addition of 2-min pulses of AA to perfused anterior pituitary cells cultured on Cytodex beads induced LH responses of short duration that coincided with the first phase of GnRH-stimulated LH release [81]. AA is formed in cultured pituitary cells stimulated by GnRH [78]. AA could be formed by monoacylglycerol and DG lipase from DG released as a result of phosphoinositide turnover. It could also arise from activation of phospholipase A_2. It seems that both enzymes are active in AA generation [82]. The site of AA action and whether it depends on Ca^{2+} or PKC for its production and action is not yet clear and awaits further investigation.

Gonadotropin Biosynthesis

The pituitary hormones, LH, FSH and thyrotropin are glycoproteins composed of 2 subunits designated α and β. The 2 subunits are bound together in noncovalent association. Within a given species the α-subunit is identical. Furthermore, the amino acid sequence of the α-subunit has been conserved in all mammalian species [83]. The β-subunit, however, is heterogenous and confers biological specifity to each hormone [83]. The 2 subunits are translated from separate mRNAs [84]. Although it is generally agreed that GnRH together with gonadal steroids regulate LH release, relatively little is known about the regulation of gonadotropin biosynthesis. Using pituitaries from ovariectomized rats, prelabeled with [^{3}H]-glucosamine and [^{14}C]-alanine, it has been shown that GnRH stimulated LH glycosylation with no effect on subunit polypeptide synthesis [85–87]. On the other hand, using much higher specific activity labeling, namely [^{35}S]-methionine as the radioactive precursor, it was found that the stimulatory effect of GnRH on polypeptide synthesis is actinomycin-D-sensitive, is detected after a lag of about 1–2 h and is mimicked by cAMP and phorbol ester in an additive manner [88–92]. The effect of gonadal steroids on LH biosynthesis is evident by the observation of a postcastration rise in gonadotropin synthesis [93, 94]. It was also reported that treatment of long-term ovariectomized rats with estradiol enhanced LH glycosylation in response to GnRH in vitro [86], and lowers the physiological concentrations of GnRH necessary to stimulate LH polypeptide synthesis, in vitro [89].

The details of the glycoprotein gene structure are beginning to emerge. The cDNAs for common α- and LH β-subunits had been cloned and sequenced in several species [95–98]. Multiple regulatory sequences of the α-gene were identified [99]. The enhancer of this gene contains 2 cAMP-responsive elements [100] that interact synergistically to increase basal and cAMP-stimulated transcription [101]. A 5′ flanking region of the LH β-gene can bind to the estrogen receptor [102] and also contains cAMP-responsive elements [103].

Recent approaches have included analysis of GnRH action on common α- and LH β-mRNAs at the pretranslational level in pituitary glands. There are several models that have been used to examine the role of GnRH deficiency on gonadotropin synthesis. In hypogonadal mice, the levels of cytosolic common α- and LH β-mRNAs are reduced to 45 and 10%, respectively, and can be restored by pulsatile GnRH treatment [104]. Similar results were obtained in hypothalamo-pituitary-disconnected ewes [105]. Pulsatile GnRH treatment of castrated, testosterone-replaced male rats resulted in a 2- to 2.5 fold increase in both LH subunit mRNA steady-state levels [106, 107]. Using the same animal model it has been demonstrated that specific frequencies and magnitudes of pulses are required to induce optimal stimulation of α- and LH β-mRNA accumulation [107–110]. These levels are normally increased during the morning of diestrus when pulsatile GnRH appears to increase [111]. A dose-dependent differential effect of GnRH on the rate and magnitude of subunit mRNA levels in rats and ewes was observed [106, 107, 109]. It was concluded that LH β-mRNA may be a limiting factor in GnRH-stimulated LH biosynthesis. In intact rats treated with GnRH antagonist, the LH β-mRNA levels decreased (57%) while α-subunit mRNA was unaffected [112]. Furthermore, in GnRH-desensitized intact rats, LH β-mRNA levels are decreased while α-mRNA is increased [113], suggesting that LH β-subunit is more tightly coupled to GnRH action. Treatment of adult rats with GnRH agonist increased the accumulation of mRNA encoding the α-subunit while LH β-mRNA levels remained unchanged [114]. The data may explain the paradoxical findings in patients treated with GnRH agonists where high plasma levels of free α-subunit were found although the pituitary was desensitized.

Several studies have examined the effect of gonadal steroids on gonadotropin gene expression. Direct cDNA hypridization analyses have shown that both LH β- and common α-subunit mRNA levels increase after castration [115–118] and decrease after steroid replacement [114, 116–119]. This postcastration effect was blocked by continuous administration of GnRH antagonist to rats or GnRH antiserum to mice and rats [112, 120].

In vitro experiments utilizing primary pituitary cell cultures are more difficult to interpret since controversy exists concerning the regulation of LH β-mRNA by GnRH. Addition of GnRH to cultured pituitary cells increased the LH β-mRNA level by 7-fold [121]. The authors suggested that this stimulatory effect is mediated by PKC, since it was reduced in PKC-depleted cells [121]. In contrast, cultures from adult female rats incubated with GnRH or GnRH agonist responded with increased α-subunit mRNA levels (2- to 5-fold), while LH β-remained unchanged and the LH response was abolished [122]. In another report it was shown that GnRH induces a rapid (detectable whithin 2 h) increase in both α- (2.5-fold) and LH β- (1.7-fold) mRNAs. The effect was reproduced by cAMP and TPA [123]. Using transcription run-on and S-1 nuclease protection assays it was recently demonstrated that GnRH-stimulated LH release is not coupled either to significant changes in the rate of LH β-gene transcription or to the cytosolic mRNA levels. However, GnRH affected the level of LH β-primary transcripts and processing intermediates, and a rapid decrease in the level of fully processed LH β mRNA in the nucleus was observed [124].

Conclusions

GnRH-induced LH synthesis and secretion is a multistep process (fig. 4). GnRH binds to a specific receptor on pituitary gonadotrophs. A 60-kilodalton membrane protein was identified as the GnRH receptor. GnRH up- and down-regulates its own receptors. The hormone-receptor complex aggregates and is later internalized for degradation of the ligand and recycling of the receptor. GnRH activates PLC most likely through the coupling of GTP-binding protein (Gp). PLC catalyzes a phospho-

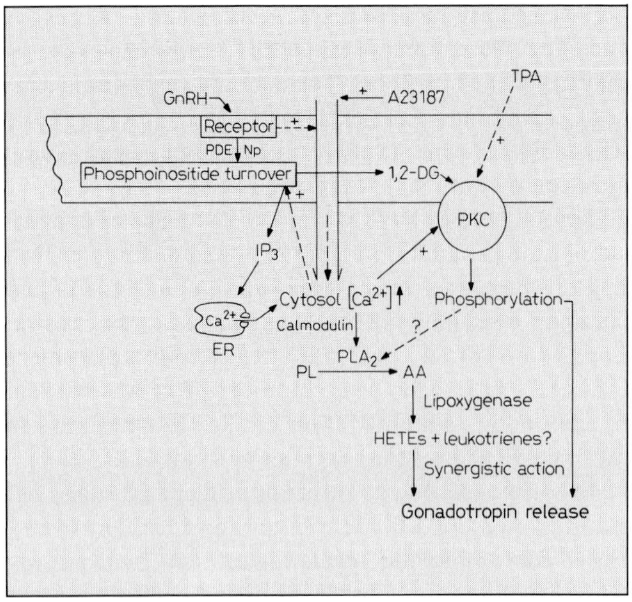

Fig. 4. Proposed model for the mechanism of action of GnRH on LH secretion and LH gene expression. The activated PKC is involved in LH gene expression. PDE = Phosphodiesterase; Np (or Gp) = a GTP-binding protein; ER = endoplasmic reticulum; PLA_2 = phospholipase A_2; PL = phospholipids; HETEs = hydroxyeicosatetraenoic acids.

diesteric hydrolysis of phosphoinositides to form IP_3 and DG. IP_3 will enhance Ca^{2+} release from intracellular stores which are probably responsible for the burst phase of LH release. At the same time the elevated $[Ca^{2+}]_i$ and DG activate PKC which phosphorylates substrate proteins to promote secretion and LH gene expression. In parallel, $[Ca^{2+}]_i$ is further elevated by Ca^{2+} influx via nifedipine-sensitive and -insensitive channels. Both Ca^{2+} and PKC participate in the sustained phase of gonadotropin release. AA, which is released at the time of GnRH stimulation, participates also with its metabolites (lipoxygenase and/or epoxygenase products) in LH exocytosis in as yet unidentified site. GnRH also regulates LH biosynthesis, and affects both protein synthesis and glycosylation. The signaling mechanism involved in GnRH-induced gonadotropin gene expression is yet unknown.

Acknowledgements

The active collaboration of Dr. J. Hermon and Dr. R. Limor is greatly appreciated. The studies were supported by the United States-Israel Binational Science Foundation and by the German-Israeli Foundation for Research and Development.

References

1 Clayton RN, Shakespear RA, Duncan JA, et al: Radio-iodinated non-degradable gonadotropin releasing hormone analogs: New probes for the investigation of pituitary gonadotropin releasing hormone receptors. Endocrinology 1979;105:1369–1381.
2 Clayton RN, Catt KJ: Gonadotropin releasing hormone receptors: Characterization, physiological regulation, and relationship to reproductive function. Endocr Rev 1981;2:186–209.
3 Naor Z, Clayton RN, Catt KJ: Characterization of GnRH receptors in cultured rat pituitary cells. Endocrinology 1980;107:1144–1152.
4 Loumaye E, Naor Z, Catt KJ: Binding affinity and biological activity of gonadotropin releasing hormone agonists in isolated pituitary cells. Endocrinology 1982;111:730–736.
5 Loumaye E, Catt KJ: Homologous regulation of gonadotropin-releasing hormone receptors in cultured pituitary cells. Science 1982;215:983–985.
6 Loumaye E, Catt KJ: Agonist induced regulation of pituitary receptors for gonadotropin-releasing hormone: Dissociation of receptor recruitment from hormone release in cultured gonadotrophs. J Biol Chem 1983;258:12002–12009.
7 Zilberstein M, Zakut H, Naor Z: Coincidence of down-regulation and desensitization in pituitary gonadotrophs by gonadotropin releasing hormone. Life Sci 1983;32:663–669.
8 Hazum E, Conn PM: Molecular mechanism of gonadotropin releasing hormone (GnRH) action. I. The GnRH receptor. Endocr Rev 1988;9:379–386.
9 Savoy-Moore RT, Schwartz NB, Duncan J, et al: Pituitary gonadotropin releasing hormone receptors during the rat estrous cycle. Science 1980;209:942–944.
10 Barkan AL, Duncan JA, Schiff M, et al: Opioid and neurotransmitter regulation of pituitary gonadotropin releasing hormone (GnRH) receptors in the ovariectomized estradiol treated rat: Role of altered GnRH secretion. Endocrinology 1985;116:1003–1010.
11 Naor Z, Schvartz I, Hazum E, et al: Effect of phorbol ester on stimulus-secretion coupling mechanisms in gonadotropin releasing hormone-stimulated pituitary gonadotrophs. Biochem Biophys Res Commun 1987;148:1312–1322.
12 Huckle WR, McArdle CA, Conn PM: Differential sensitivity of agonist- and antagonist-occupied gonadotropin-releasing hormone receptors to protein kinase C activators. J Biol Chem 1988;263:3296–3302.
13 Hirota K, Hirota T, Aguilera G, et al: Hormone induced redistribution of calcium activated phospholipid dependent protein kinase in pituitary gonadotrophs. J Biol Chem 1985;260:3243–3246.
14 Naor Z, Zer J, Zakut H, et al: Characterization of pituitary calcium-activated phospholipid-dependent protein kinase: Redistribution by gonadotropin releasing hormone. Proc Natl Acad Sci USA 1985;82:8203–8207.
15 Hazum E: Photoaffinity labeling of luteinizing hormone releasing hormone receptor of rat pituitary membrane preparations. Endocrinology 1981;109:1281–1284.
16 Iwashita M, Catt KJ: Photoaffinity labeling of pituitary and gonadal receptors for gonadotropin-releasing hormone. Endocrinology 1985;117:738–746.
17 Eidne KA, Hendricks DT, Millar RP: Demonstration of a 60K molecular weight luteinizing hormone-releasing hormone recep-

tor in solubilized adrenal membrane by a ligand-immunoblotting technique. Endocrinology 1985;116:1792–1795.
18 Hsueh AJW, Jones PBC: Extrapituitary actions of gonadotropin releasing hormone. Endocr Rev 1981;2:437–461.
19 Knecht M, Rahta T, Naor Z, et al. Direct effects of GnRH on the ovary; in Greenwald GS, Terranova PF (eds): Factors Regulating Ovarian Functions. New York Raven Press, 1983, pp 225–243.
20 Naor Z, Dekel N, Molcho J, et al: Phospholipid turnover as a mode of gonadotropin-releasing hormone action in the ovary: in Runnebaum B, Rabe T, Kiesel L, Merz WE (eds): Secretion and Action of Gonadotropins, Physiology and Clinic. Berlin, Springer, 1984, pp 83–94.
21 Naor Z, Molcho J, Zilberstein M, et al: Phospholipid turnover in gonadotropin releasing hormone target cells: Comparative studies; in McKerns KW, Naor Z (eds): Hormonal Control of the Hypothalamo-Pituitary-Gonadal Axis. New York Plenum Press, 1984, pp 493–508.
22 Birnbaumer L, Shahabi N, Rivier J, et al: Evidence for a physiological role of gonadotropin releasing hormone (GnRH) or GnRH-like material in the ovary. Endocrinology 1985;116: 1367–1370.
23 Hazum E, Schvartz I, Waksman Y, et al: Solubilization and purification of rat pituitary gonadotropin-releasing hormone receptor. J Biol Chem 1986;261:13043–13048.
24 Iwashita M, Hirota K, Izumi SI, et al: Solubilization and characterization of the rat pituitary gonadotropin releasing hormone receptor. J Mol Endocrinol 1988;1:187–196.
25 Hazum E, Schvartz I, Popliker M: Production and characterization of antibodies to gonadotropin-releasing hormone receptor. J Biol Chem 1987;262:531–534.
26 Naor Z, Childs GV: Binding and activation of gonadotropin-releasing hormone receptors in pituitary and gonadal cells. Int Rev Cytol 1986;103:147–187.
27 Naor Z, Childs GV, Leifer AM, et al: Gonadotropin releasing hormone binding and activation of enriched population of pituitary gonadotrophs. Mol Cell Endocrinol 1982;25:85–97.
28 Hyde CL, Childs (Moriaty) G, Wahl LM, et al: Preparation of gonadotroph enriched cell population from adult rat anterior pituitary cells by centrifugal elutriation. Endocrinology 1982; 111:1421–1423.
29 Childs GV, Naor Z, Hazum E, et al: Cytochemical characterization of pituitary target cells for biotinylated gonadotropin releasing hormone. Peptides 1983;4:549–555.
30 Hazum E, Cuatrecasas P, Marian J, et al: Receptor-mediated internalization of gonadotropin-releasing hormone by pituitary gonadotropes. Proc Natl Acad Sci USA 1980;77:6692–6695.
31 Naor Z, Atlas D, Clayton RN, et al: Interaction of fluorescent gonadotropin-releasing hormone with receptors in cultured pituitary cells. J Biol Chem 1981;256:3049–3052.
32 Childs GV, Hyde C, Naor Z, et al: Heterogeneous luteinizing hormone and follicle stimulating hormone storage patterns in subtypes of gonadotropes separated by centrifugal elutriation. Endocrinology 1983;113:2120–2128.
33 Sternberger LA, Petrali JP: Quantitative immunocytochemistry of pituitary receptors for luteinizing hormone-releasing hormones. Cell Tissue Res 1975;162:141–176.
34 Sternberger LA, Petrali JP, Joseph SA, et al. Specificity of the immunocytochemical luteinizing hormone-releasing hormone receptor reaction. Endocrinology 1978;102:63–73.
35 Childs GV: Application of dual pre-embedding stains for gonadotropins to pituitary cell monolayers with avidin-biotin (ABC) and peroxidase-antiperoxidase (PAP) complexes: Light microscopic studies. Stain Technol 1984;58:281–290.
36 Schrey MP: Gonadotropin releasing hormone stimulates the formation of inositol phosphates in rat anterior pituitary tissue. Biochem J 1985;226:563–569.
37 Naor Z, Azrad A, Limor R, et al: Gonadotropin releasing hormone activates a rapid Ca^{2+}-independent phosphodiester hydrolysis of polyphosphoinositides in pituitary gonadotrophs. J Biol Chem 1986;261:12506–12512.
38 Morgan RO, Chang JP, Catt KJ: Novel aspects of gonadotropin releasing hormone action on inositol phosphate metabolism in cultured pituitary gonadotrophs. J Biol Chem 1987;262:1166–1171.
39 Nishizuka Y: The role of protein kinase C in cell surface signal transduction and tumor protomotion. Nature 1984;308:693–698.
40 Nishizuka Y: Studies and perspectives of protein kinase C. Science 1986;233:305–312.
41 Berridge MJ, Irvine RF: Inositol trisphosphate, a novel second messenger in cellular signal transduction. Nature 1984;312:315–321.
42 Limor R, Ayalon D, Capponi AM, et al: Cytosolic free Ca^{2+} levels in cultured pituitary cells separated by centrifugal elutriation: Effect of gonadotropin releasing hormone. Endocrinology 1987; 120:497–503.
43 Irvine RF, Moore RM: Microinjection of inositol 1,3,4,5-tetrakisphosphate activates sea urchin eggs by a mechanism dependent on external Ca^{2+}. Biochem J 1986;240:917–920.
44 Morris AP, Gallacher DV, Irvine RF, et al: Synergism of inositol trisphosphate and tetrakisphosphate activating Ca^{2+} dependent K^{+} channels. Nature 1987;330:653–655.
45 Majerus PW, Connolly TM, Bansal VS, et al: Inositol phosphates: synthesis and degradation. J Biol Chem 1988;263:3051–3054.
46 Catt KJ, Chang JP, Stojilkovic S, et al: GnRH receptors and activation of gonadotropin secretion; in Lizuka R, Semm KK (eds): Human Reproduction: Current Status/Future Prospects. Amsterdam, Elsevier, 1988, pp 37 55.
47 Gilman AG: G proteins: Transducers of receptor generated signals. Annu Rev Biochem 1987;56:615–649.
48 Martin TFJ, Lucas DO, Baggalieh SM, et al: Thyrotropin-releasing hormone activates a Ca^{2+} dependent polyphosphoinositide phosphodiesterase in permeable GH_3 cells. J Biol Chem 1986; 261:2918–2927.
49 Cockcroft S, Gomperts BD: Role of guanine nucleotide binding protein in the activation of polyphosphoinositide phosphodiesterase. Nature 1985;314:534–536.
50 Limor R, Schvartz I, Hazum E, et al: Effect of guanine nucleotides on phospholipase C activity in permeabilized pituitary cells: Possible involvement of an inhibitory GTP-binding protein. Biochem Biophys Res Commun 1989;159:209–215.
51 Conn PM, Marian J, McMillan M, et al: Gonadotropin releasing hormone action in the pituitary: A three step mechanism. Endocr Rev 1981;2:174–185.
52 Shangold GA, Murphy SN, Miller RJ: Gonadotropin-releasing hormone induced Ca^{2+} transients in single identified gonadotropes require both intracellular Ca^{2+} mobilization and Ca^{2+} influx. Proc Natl Acad Sci USA 1988;85:6566–6570.

53 Naor Z, Capponi AM, Rossier MF, et al: Gonadotropin releasing hormone-induced rise in cytosolic free Ca^{2+} levels. Mobilization of cellular and extracellular Ca^{2+} pools and relationship to gonadotropin secretion. Mol Endocrinol 1988;2:512–520.
54 Davidson JS, Wakefield IK, King JA, et al: Dual pathways of calcium entry in spike and plateau phase of luteinizing hormone release from chicken pituitary cells: Sequential activation of receptor-operated and voltage-sensitive Ca^{2+} channels by gonadotropin-releasing hormone. Mol Endocrinol 1988;2:382–390.
55 Conn PM: The molecular basis of gonadotropin releasing hormone action. Endocr Rev 1986;7:3–10.
56 Chang JP, McCoy EE, Graeter J, et al: Participation of voltage-dependent calcium channels in the action of gonadotropin releasing hormone. J Biol Chem 1986;261:9105–9108.
57 Mason NT, Waring DN: Patch-clamp recording of single ion channel activation by gonadotropin-releasing hormone in ovine pituitary gonadotrophs. Neuroendocrinology 1986;43:205–219.
58 Woods NM, Cuthbertson KSR, Cobbold PH: Repetitive transient rises in cytoplasmic free calcium in hormone stimulated hepatocytes. Nature 1988;319:600–602.
59 Berridge MJ, Cobbold PH, Cuthbertson KSR: Spatial and temporal aspects of cell signalling. Phil Trans R Soc Lond [B] 1988; 320:325–343.
60 Turgeon JL, Ashcroft SJH, Waring DW, et al: Characteristics of adenohypophyseal Ca^{2+}-phospholipid dependent protein kinase. Mol Cell Endocrinol 1984;34:107–112.
61 Hermon J, Reiss N, Naor Z: Phospholipid dependent Ca^{2+} activated protein kinase (C-kinase) in the pituitary: Further characterization and endogenous redistribution. Mol Cell Endocrinol 1986;47:201–208.
62 Fearon CW, Tashjian AH Jr: Thyrotropin-releasing hormone induces redistribution of protein-kinase C in GH_4C_1 rat pituitary cells. J Biol Chem 1985;260:8366–8371.
63 Kraft AS, Anderson WB: Phorbol ester increases the amount of Ca^{2+} phospholipid-dependent protein kinase associated with plasma membrane. Nature 1983;301:621–623.
64 Naor Z, Eli Y: Synergistic stimulation of luteinizing hormone release by protein kinase C activators and Ca^{2+} ionophore. Biochem Biophys Res Commun 1985;130:848–853.
65 Nishizuka Y: The molecular heterogeneity of protein kinase C and its implications for cellular regulation. Nature 1988;334: 661–665.
66 Parker PJ, Coussens L, Totty N, et al: The complete primary structure of protein kinase C – the major phorbol ester receptor. Science 1986;233:853–859.
67 Kikkawa U, Uno Y, Ogita K, et al: Identification of the structures of multiple subspecies of protein kinase C expressed in rat brain. FEBS Lett 1987;217:227–231.
68 Ono Y, Kikkawa U, Ogita K, et al: Expression and properties of two types of protein kinase C determined by alternative splicing from a single gene. Science 1987;236:1116–1120.
69 Naor Z, Dan-Cohen H, Hermon J, et al: Induction of exocytosis in permeabilized pituitary cells by α- and β-type protein kinase C. Proc Natl Acad Sci USA 1989;88:4501–4504.
70 Naor Z: Further characterization of protein kinase C subspecies in the hypothalamopituitary axis: Differential activation by phorbol ester. Endocrinology 1990;126:1521–1526.
71 Dan Cohen H, Naor Z: Mechanism of action of gonadotropin releasing hormone upon gonadotropin secretion: Involvement of protein kinase C as revealed by staurosporine inhibition and enzyme depletion. Mol Cell Endocrinol 1990;69:135–144.
72 McArdle CA, Huckle WR, Conn PM: Phorbol ester reduces gonadotropes responsiveness to protein kinase activators but not to Ca^{2+} mobilizing secretagogues. J Biol Chem 1987;262:5028–5035.
73 Stojilkovic SS, Chang JP, Ngo D, et al: Evidence for a role of protein kinase C in luteinizing hormone secretion. J Biol Chem 1988;263:17307–17311.
74 Strulovici B, Daniel-Issakani S, Oto E, et al: Activation of distinct protein kinase C isozymes by phorbol esters: Correlation with induction of interleukin I gene expression. Biochemistry 1989;28:3569–3576.
75 Huang FL, Yoshida Y, Cunha-Melo JR: Differential down-regulation of protein kinase C isozymes. J Biol Chem 1989;264: 4238–4243.
76 Fournier A, Hardy SJ, Clark KY, et al: Phorbol ester induces differential membrane-association of protein kinase C subspecies in human platelets. Biochem Biophys Res Commun 1989; 161:556–561.
77 Katsuhiko A, Nicola B, Kikkawa U, et al: Differential down-regulation of protein kinase C subspecies in KM3 cells. FEBS Lett 1988;236:396–400.
78 Naor Z, Kiesel J, Vanderhoek J, et al: Mechanism of action of gonadotropin releasing hormone, role of lipoxygenase products of arachidonic acid in luteinizing hormone release. J Steroid Biochem 1985;23:711–717.
79 Kiesel L, Przylipiak A, Rabe T, et al: Leukotrienes stimulate gonadotropin release in vitro. Gynecol Endocrinol 1987;1:25–35.
80 Snyder GD, Capdevila NC, Nanna S, et al: Action of luteinizing hormone-releasing hormone: Involvement of novel arachidonic acid metabolites. Proc Natl Acad Sci USA 1983;80:3504–3507.
81 Chang PC, Graeter J, Catt KJ: Dynamic action of arachidonic acid and protein kinase C in pituitary stimulation by gonadotropin releasing hormone. Endocrinology 1987;120:1837–1845.
82 Chang JP, Morgan RO, Catt KJ: Dependence of secretory responses to gonadotropin-releasing hormone on diacylglycerol metabolism. J Biol Chem 1988;263:181614–18620.
83 Pierce JG, Parsons JF: Glycoproteins hormones: Structure and function. Annu Rev Biochem 1980;50:465–495.
84 Chin WW: Glycoprotein hormone genes; in Habener JF (ed): Molecular Cloning of Hormone Genes. Clifton, Humana Press, 1987, pp 137–172.
85 Liu T-C, Jackson GL, Gorski J: Effects of synthetic gonadotropin releasing hormone on incorporation of radioactive glucoseamine and amino acids into luteinizing hormone and total protein by rat pituitaries in vitro. Endocrinoloy 1976;98:151–163.
86 Liu T-C, Jackson GL: Effect of in vivo treatment with estrogen on luteinizing hormone synthesis and release by rat pituitaries in vitro. Endocrinology 1977;100:1294–1302.
87 Liu T-C, Jackson GL: Modifications of luteinizing hormone biosynthesis and release by gonadotropin releasing hormone, and actinomycin D. Endocrinology 1978;103:1253–1263.
88 Khar A, Debebyuk L, Jutisz M: Biosynthesis of gonadotropin by rat pituitary cells in culture and in pituitaries homogenates: Effect of gonadotropin releasing hormone. Mol Cell Endocrinol 1978;12:53–65.

89 Ramey JW, Highsmith RE, Wilfinger WW, et al: the effect of gonadotropin releasing hormone and estradiol on luteinizing hormone biosynthesis in cultured rat anterior pituitary cells. Endocrinology 1987;120:1503–1513.
90 Starzec A, Jutisz M, Counis R: Cyclic adenosine monophosphate and phorbol ester, like gonadotropin-releasing hormone, stimulates the biosynthesis of luteinizing hormone polypeptide chains in nonadditive manner. Mol Endocrinol 1989;3:618–624.
91 Starzec A, Counis R, Jutisz M: Gonadotropin releasing hormone stimulates the synthesis of the polypeptide chains of luteinizing hormone. Endocrinology 1986;119:561–565.
92 Starzec A, Jutisz M, Counis R: Detection of a lag phase in gonadotropin-releasing hormone stimulated synthesis of lutropin peptide chains in cultured rat anterior pituitary cells. Biochem Biophys Res Commun 1988;153:664–670.
93 Godine JE, Chin WW, Habener JF: Luteinizing and follicle stimulating hormones. J Biol Chem 1980;255:8780–8783.
94 Counis R, Corbani M, Jutisz M: Estradiol regulates mRNAs encoding precursors of lutropin (LH) subunits. Biochem Biophys Res Commun 1983;114:65–72.
95 Nilson JH, Thomason AR, Cserback MT, et al: Nucleotide sequence of a cDNA for the common subunit of the bovine pituitary glycoprotein hormones. J Biol Chem 1983;258:4679–4682.
96 Chin WW, Godine JE, Klein Dr, et al: Nucleotide sequence of the cDNA encoding the precursor of the β subunit of the rat lutropin. Proc Natl Acad Sci USA 1983;80:4649–4653.
97 Jameson L, Chin WW, Hollenberg AW, et al: The gene encoding the β subunit rat luteinizing hormone. J Biol Chem 1984; 259:15474–15480.
98 Kato Y, Hirai T: Cloning and DNA sequence analysis of the cDNA for the precursor of porcine luteinizing hormone (LH) β subunit. Mol Cell Endocrinol 1989;62:47–53.
99 Franklyn JA, Balfour NJ, Gurr JA: Cis acting elements of the human common α subunit gene determine expression in pituitary cells (abstract 78). Proc 71st Annu Meet Am Endocr Soc, Seattle, 1989, p 42.
100 Delegeane AM, Ferland LH, Mellon PL: Tissue-specific enhancer of the human glycoprotein hormone α-subunit gene: Dependence on cyclic AMP-inducible elements. Mol Cell Biol 1987;7:3994–4002.
101 Andersen B, Bokar JA, Kennedy G, et al: Spacing between the two CREs in the alpha gonadotropin gene decreases transcription and binding affinity of transacting factors for CREs (abstract 80). Proc 71st Annu Meet Am Endocr Soc, Seattle, 1989, p 42.
102 Shupnik MA, Weinmann CM, Notides AC, et al: An upstream region of the rat luteinizing hormone β gene binds estrogen receptor and confers estrogen responsiveness. J Biol Chem 1989;264:80–86.
103 Clayton RN, Powell J, Lalloz MRA, et al: The rat LH-beta gene contains a cAMP responsive element (abstract 77). Proc 71st Annu Meet Am Endocr Soc, Seattle, 1989, p. 42.
104 Saade G, London DR, Clayton RN: The interaction of gonadotropin-releasing hormone and estradiol on luteinizing hormone and prolactin gene expression in female hypogonadal (hpg) mice. Endocrinology 1989;124:1744–1755.
105 Mercer JE, Clements JA, Funder JW, et al: Luteinizing hormone-β-mRNA levels are regulated primarily by gonadotropin releasing hormone and not by negative estrogen feedback in the pituitary. Neuroendocrinology 1988;47:563–566.
106 Papavasiliou SS; Zmeili S, Khoury S, et al: Gonadotropin-releasing hormone differentially regulates expression of the genes for luteinizing hormone α and β subunits in male rats. Proc Natl Acad Sci USA 1986;83:4026–4029.
107 Haisenleder DJ, Khoury S, Zmeili SM, et al: The frequency of gonadotropin releasing hormone secretion regulates expression of luteinizing hormone β-subunit messenger ribonucleic acids in male rats. Mol Endocrinol 1987;1:834–838.
108 Leung K, Kaynard AH, Negrini BP, et al: Differential regulation of gonadotropin subunit messenger ribonucleic acids by gonadotropin releasing hormone pulse frequency in ewes. Mol Endocrinol 1987;1:724–728.
109 Dalkin AC, Haisenleder DJ, Ortolano GA, et al: The frequency of gonadotropin-releasing hormone stimulation differentially regulates gonadotropin subunit messenger ribonucleic acid expression. Endocrinology 1989;125:917–924.
110 Haisenleder DJ, Katt JA, Ortolano GA, et al: Influence of gonadotropin releasing hormone pulse amplitude, frequency, and treatment duration of the regulation of luteinizing hormone (LH) subunit messenger ribonucleic acids and LH secretion. Mol Endocrinol 1988;2:338–343.
111 Zmeili SM, Papavasiliou SS, Thorner MO, et al: Alpha and luteinizing hormone beta subunit messenger ribonucleic acids during the rat estrous cycle. Endocrinology 1986;119:1867–1869.
112 Lalloz MRA, Detta A, Clayton RN: Gonadotropin releasing hormone is required for enhanced luteinizing hormone subunit gene expression in vivo. Endocrinology 1988;122:1681–1688.
113 Lalloz MRA, Detta A, Clayton RN: Gonadotropin-releasing hormone desensitization preferentially inhibits expression of the luteinizing hormone β-subunit gene in vivo. Endocrinology 1988;122:1689–1694.
114 Simard J, Labrie C, Hubert JF, et al: Modulation by sex steroids and [D-Trp^6, Des-Gly-NH_2^{10}]luteinizing hormone (LH)-releasing hormone ethylamide of α-subunit and LH β-messenger ribonucleic acid levels in the rat anterior pituitary gland. Mol Endocrinol 1988;2:775–784.
115 Abbot SD, Docherty K, Roberts JL, et al: Castration increases luteinizing hormone subunit messenger RNA levels in male rat pituitaries. J Endocrinol 1985;107:R1–R4.
116 Papavasiliou SS, Zmeili S, Herbon L, et al: α-and luteinizing hormone β messenger ribonucleic acid (RNA) of male and female rats after castration: Quantitation using an optimized RNA dot blot hybridization assay. Endocrinology 1986;119: 691–698.
117 Gharib SD, Bowers SM, Reed LR, et al: Regulation of rat luteinizing hormone subunit messenger ribonucleic acids by gonadal steroids hormones. J Clin Invest 1986;77:582–589.
118 Nilson JH, Neyidlik MT, Virgin JB, et al: Expression of α subunit and luteinizing hormone β genes in the ovine anterior pituitary: Estradiol suppresses accumulation of mRNAs for both α and β luteinizing hormone. J Biol Chem 1983;258: 12087–12090.
119 Landefeld TD, Kepa J, Karsch FJ: Regulation of subunit synthesis by gonadal steroid feedback in the sheep anterior pituitary. J Biol Chem 1983;258:2390–2393.
120 Kato Y, Imar K, Sakai T, et al: Simultaneous effect of gonadotropin-releasing hormone (GnRH) on the expression of two

gonadotropin β genes by passive immunization to GnRH. Mol Cell Endocrinol 1988;62:135–139.
121 Andrews WV, Maurer RA, Conn PM: Stimulation of rat luteinizing hormone-β messenger RNA levels by gonadotropin releasing hormone. J Biol Chem 1988;263:13755–13761.
122 Hubert JF, Simard Y, Gagne B, et al: Effect of luteinizing hormone releasing hormone (LHRH) and [*D*-Trp^6, Des-Gly-NH_2^{10}]LHRH ethylamide on α subunit and LH β messenger ribonucleic acid levels in rat anterior pituitary cells in culture. Mol Endocrinol 1988;2:521–527.
123 Starzec A, Moumni M, d'Angelo-Bernard G, et al: Activation of protein kinases A and C, like stimulation of gonadotropin-releasing hormone, enhances levels of both alpha and LH beta mRNAs in cultured pituitary cells (abstract 242). Proc 71st Annu Meet Endocr Soc. Seattle, 1989.
124 Salton SRJ, Blum M, Jonassen JA, et al: Stimulation of pituitary luteinizing hormone secretion by gonadotropin releasing hormone is not coupled to β-luteinizing hormone gene transcription. Mol Endocrinol 1988;2:1033–1042.

Prof. Z. Naor
Department of Biochemistry
George S. Wise Faculty of Life Sciences
Tel Aviv University
Tel Aviv 69978 (Israel)

Discussion

M. Aubert, Switzerland: I would like to congratulate you on your excellent presentation. My question refers to the work performed by Dr. Eli Hazum, who purified GnRH receptor proteins and was able to raise antibodies against this protein. Did you try to use such antiserum to localize GnRH receptors in the pituitary or in other tissues and for the purification of GnRH receptors?

Z. Naor, Israel: We use a different technique in our trials to follow the fate of the hormone-receptor complex: we use a *D*-lys^6-GnRH-biotin probe.

Horm Res 1990;33:87–94

0301–0163/90/0334–0087$2.75/0

Androgen Receptor Defects in Patients with Minimal and Partial Androgen Resistance Classified According to a Model of Androgen-Receptor Complex Energy States

M. Kaufman[a], *L. Pinsky*[a], *B. Gottlieb*[a], *M. Schweitzer*[a], *A. Brezezinski*[a], *C. von Westarp*[b], *J. Ginsberg*[b]

[a] Lady Davis Institute for Medical Research of the Sir Mortimer B. Davis-Jewish General Hospital, Centre for Human Genetics, McGill University, Montreal; [b] Division of Endocrinology and Metabolism, University of Alberta, Edmonton, Canada

Key Words. Androgen receptor · Androgen resistance

Abstract. We have characterized intracellularly the androgen-receptor (A-R) complexes formed by genital skin fibroblasts from 2 unrelated males with qualitative defects of the androgen receptor: one has a small nonhypospadic penis as part of a syndrome of mild androgen resistance; the other was born with ambiguous external genitalia. The dissociation rate constants of testosterone, methyltrienolone (MT), dihydrotestosterone (DHT) and mibolerone (MB) from normal androgen receptors were determined at various temperatures: when plotted by the method of Arrhenius, they yielded a linear hierarchy of dissociation states with energies of state IV > III > II > I, respectively. Relative to this hierarchy, patient A-R complexes were displaced to higher, androgen-inappropriate energies in a mutant-distinctive pattern. MB- or MT-R complexes of both patients were thermolabile; however, both up-regulated normally in response to prolonged incubation with either hormone. Apparent equilibrium affinity constants (K_d) of the DHT- and MB-R complexes formed by both patients were normal; however, the binding capacity (B_{max}) for MB in 1 case was subnormal. The distinctive biochemical phenotypes of A-R complexes in these 2 patients with androgen resistance will facilitate the definition of structure-function relations in the androgen receptor, a classical DNA-binding, transcription-regulating protein.

Introduction

The essential role of the androgen receptor in mediating the diverse morphogenetic and physiological effects of androgens is underscored by androgen resistance (insensitivity) syndromes that result from quantitative or qualitative defects in this molecule. Patients with these syndromes are classified as 'complete' if their external genitalia are female, 'minimal' if they are male and 'partial' if they are morphologically ambivalent to any substantive degree. Qualitative receptor defects have been identified in patients with phenotypic abnormalities that encompass the entire spectrum of androgen resistance syndromes. The list of defects that has been recognized within genital skin fibroblasts (GSFs) includes: increased thermolability of androgen-receptor (A-R) complexes [1, 2], failure of up-regulation of the androgen receptor by androgens [3, 4], decreased equilibrium affinities (K_d) of the A-R complexes [5, 6] and increased nonequilibrium rate constants of dissociation of various ligands from the androgen receptor [7, 8]. The latter property has been particularly informative because it has allowed us to recognize 2 classes of qualitative receptor abnormalities: ligand-selective [9, 10] and ligand-independent [5, 11]. In the latter the dissociative behavior of mutant androgen receptors is abnormal whether complexed with mibolerone (MB), dihydrotestosterone (DHT) or methyltrienolone (MT); in the former, rate constants of MT-R complexes are normal, but those of either MB- or DHT-R complexes are abnormal.

Here we report the use of these markers to identify qualitative receptor abnormalities in 2 additional patients – one with minimal, the other with partial, androgen resistance. We have exploited the effect of temperature on rate constants of dissociation to define a hierarchy of energy states for MB-, DHT-, MT- and testosterone (T)-R complexes in control GSFs. The dissociation constants of the patients' MB-, DHT- and MT-R complexes are shifted to higher energies in this hierarchy in a patient-distinctive fashion: A-R complexes in GSFs from the patient with ligand-selective, mild androgen resistance are less severely affected than those from the patient with ligand-independent, partial androgen resistance. We have additionally defined a process, termed 'recruitment', wherein replicate cultures of normal cells, co-incubated in the presence of cycloheximide, display a greater maximum androgen-receptor binding activity with MB compared to DHT. In 1 patient this process was defective.

Table 1. Dissociation rate constants of various A-R complexes within control GSFs

Ligand	Temperature °C	Rate constant min^{-1}	References
MB	37	0.003	1
	40	0.006[1] (n = 10)	
	42	0.008	17
DHT	29	0.002	18
	32	0.003	18
	37	0.006	18
	40	0.01	18
MT	30	0.004	7
	33	0.005	7
	37	0.012	7
	40	0.024	7
T	27	0.006	19
	32	0.014	19
	37	0.024	19
	40	0.05	19

[1] SD = ± 0.001.

Methods

Subjects

An Inuit Eskimo boy (patient No. 3287) was first examined at the Montreal Children's Hospital at 2 years, 3 months of age because of severe perineal hypospadias with chordee. His karyotype is 46,XY. The penis was 3.5 × 1.0 cm. The scrotum was fused but empty. The testes, 1.5 × 1.0 cm, were palpable at the external inguinal rings. Pelvic ultrasonography did not reveal any female genitalia. At 3 years, 5 months of age his total serum testosterone was less than 1 n*M*, and his luteinizing hormone (LH) and follicle-stimulating hormone (FSH) values were 4.8 and 7.9 IU/l, respectively. Three days after β-human chorionic gonadotropin stimulation, 1,500 units per day, his total serum testosterone was 4.2 n*M*, and his LH, FSH values were 118.5 (cross-reactivity) and 7.6 IU/l, respectively. There is nothing remarkable about the sexual development of his nonconsanguineous parents, his sister (10 years) or brother (7 years).

A Caucasian boy (patient No. 4007) was first seen at the University of Alberta Hospital at 18 years of age with the complaints of bilateral gynecomastia and subnormal virilization. His karyotype is 46,XY. There was little facial hair, no chest hair, his voice was high-pitched, and stretched penis length was 6.5 cm. The testes felt normal: they were 15 ml bilaterally. Public hair was Tanner IV. Bone age was 17 years. His father had normal secondary sexual development. His mother and sister each had menarche at age 12; we do not know whether either has sparse or asymmetrical pubic or axillary hair, or whether hair development in either area was delayed. His total plasma testosterone was 48.4 n*M* (normal: 12–35) and his LH was 37.5 IU/l (normal: 4–23). FSH, 5α-DHT, androstenedione and estradiol were normal. Sperm analysis on a semen sample of 0.7 ml revealed a density of 35 million/ml, but 60% were dead, 80% were immotile at 2.5 h, 13% were tapered and 11% were amorphous. Gynecomastia was treated with reduction mammoplasty.

He was re-evaluated at 19.5 years. He still had scant upper lip and sideburn hair, and shaved once per 2 weeks. His voice was still prepubertal. Axillary and pubic hair were normal. Chest hair was absent, and leg hair scant. His total plasma testosterone was 51 n*M* and his LH was 13.9 IU/l (normal: 2–12). His FSH was normal. An intravenous gonadotropin-releasing hormone stimulation test provoked a rise in basal LH (15.2 IU/l) to a peak 45-min value of 79.3. His basal FSH, 3.7 IU/l (normal: 1–7), had a peak 75-min value of 5.5. A repeat sperm analysis on a semen sample of 0.5 ml revealed a density of only 5 million/ml; 90% were dead, none were motile at 25 h and 17% were amorphous.

Cell Culture

The GSF strains were developed in our laboratory, or at the University of Alberta (strain 4007) from small pieces of labium majus, scrotal or preputial skin obtained with informed consent according to protocols approved by the Hospital Ethics Committee. Monolayers were grown to confluence in 5-cm^2 Petri dishes with Opti minimal essential medium (Gibco, Grand Island, N.Y., USA) supplemented with 5% fetal calf serum, 1 m*M* pyruvate, 10 mg/l garamycin and 60 mg/l each of penicillin G and streptomycin sulfate. 24 h before assay for A-R activity, this medium was removed and the monolayers were incubated with serum-free (SF) minimal essential medium (Hanks' salts) buffered to pH 7.4 with 15 m*M* HEPES. In some assays this medium contained 10 m*M* cycloheximide to prevent up-regulation of androgen receptor activity [12].

Binding of ^{3}H-MB and ^{3}H-DHT to the Androgen Receptor within Intact GSFs

Replicate monolayers in 5-cm^2 dishes were incubated in an SF medium with 10 μ*M* cycloheximide in the presence of 0.05–3.0 n*M* [17α-methyl-^{3}H]-MB (75 Ci/mmol) or [1,2,4,5,6,7-^{3}H]-DHT (147 Ci/mmol) alone (to measure total binding) or together with a 200-fold excess of the respective radioinert androgen (to measure nonspecific binding). After incubation for 2 h in a humidified 37 °C incubator supplied with 5% CO_2:95% air, the dishes were placed on a bed of ice, washed twice with 5 ml of 20 m*M* Tris-HCl (pH 7.4) containing 0.15 *M* NaCl and 0.2% bovine serum albumin (BSA) and

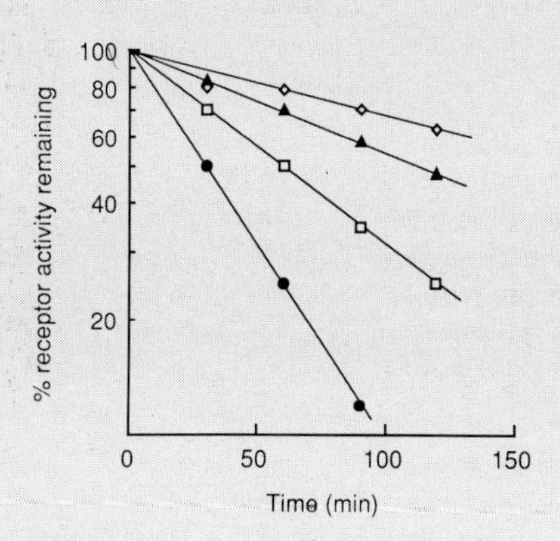

Fig. 1. Dissociation of A-R complexes within intact cells. Control cell monolayers were labeled with 3 nM ^{3}H-T (●), ^{3}H-MT (□), ^{3}H-DHT (▲) and ^{3}H-MB (◇) for 2 h at 37 °C, and the A-R complexes chased at 37 °C with a 200-fold excess of the respective radioinert androgen.

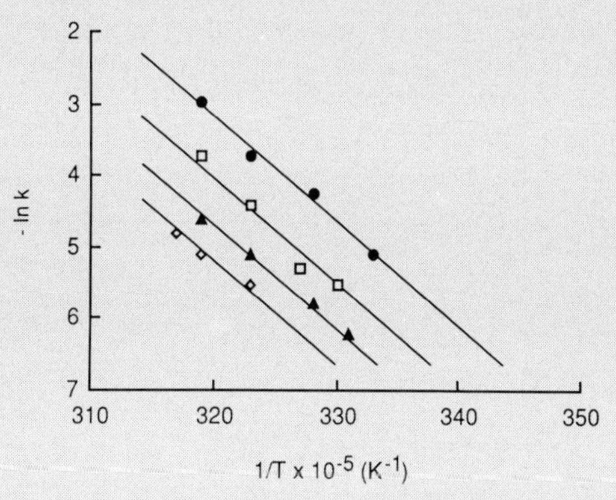

Fig. 2. Energy states of the androgen receptor. Rate constants calculated from the slopes of rate data for T-R (●), MT-R (□), DHT-R (▲) and MB-R complexes (◇) at various temperatures (table 1) plotted by the method of Arrhenius. MB-R complexes (state I) are closest to and T-R (state IV) complexes farthest from the origin.

twice with the same buffer lacking BSA. The monolayers were then treated for 5 min at room temperature with 2 ml of 0.1% trypsin. The loosely adherent cells were scraped with a rubber policeman and centrifuged (4 °C) at 2,000 g for 5 min. The cells were resuspended in Tris buffer lacking BSA and recentrifuged. The final cell pellets were solubilized with 0.5 N NaOH (1.5 ml) and sampled for protein [13] and radioactivity, the latter in 10 ml of a toluene solution containing Omnifluor (4 g/l; New England Nuclear, Boston, Mass., USA) that was counted in an LKB 1217 RACKBETA scintillation counter at an efficiency of 60%. Specific binding (total minus nonspecific) was plotted as a function of the free androgen concentration by the method of Scatchard [14] and values of B_{max} and K_d were derived from lines fitted by linear regression.

Dissociation of A-R Complexes

To determine the dissociation rate of A-R complexes within cells that had been incubated (37 °C) for 2 h with 3 nM ^{3}H-MB, ^{3}H-DHT, [17α-methyl-^{3}H]-MT (87 Ci/mmol) or [1,2,6,7-^{3}H]-T (105 Ci/mmol) as above, the assay medium in some dishes was replaced by a 'chase' medium containing 0.6 μM of the respective radioinert androgen. After various periods of time at temperatures between 20 and 42 °C, the percent A-R complexes remaining was determined and the results plotted semilogarithmically. Rate constants were calculated from slopes of regression lines fitted to these data. In all assays, the concentration of specific A-R complexes at each time was computed by subtracting the mean nonspecific binding measured in duplicate from that of total binding determined in triplicate. Arrhenius plots were constructed by plotting the natural logarithm (–ln) of the dissociation rate constant versus the reciprocal of the absolute temperature (1/T = 1/C + 273) where C = degrees Celsius. In assays with ^{3}H-T we used GSFs from patients with 5α-reductase deficiency.

Up-Regulation of A-R Complexes

To assess the ability of mutant androgen receptors to up-regulate, monolayers were incubated at 37 °C in SF media with 3 nM ^{3}H-MB or ^{3}H-MT in the presence or absence of 200-fold radioinert androgen. Specific binding activity was measured at 1, 5, 24, 48 and 72 h.

Thermostability of Pools of A-R Complexes

To assess the effect of thermal stress on stability of populations of A-R complexes within GSFs, the medium of some replicate monolayers that had been incubated at 37 °C with 3 nM ^{3}H-MB or ^{3}H-MT in the presence or absence of 200-fold radioinert androgen overnight was replaced by a medium that contained both 10 μM cycloheximide as well as a reduced (1 nM) concentration of ^{3}H-androgens. The monolayers were then postincubated at 42 °C and the amount of specific binding activity measured at intervals for up to 6 h.

Results

Rates of dissociation for control T-R complexes (within cells from patients with 5α-reductase deficiency) > MT-R complexes > DHT-R complexes > MB-R complexes (fig. 1). When plotted by the method of Arrhenius (fig. 2) the rate constants calculated from the slopes of these lines and from those determined at other temperatures (table 1) yielded a hierarchy of energy states (IV..I): MB-R complexes (state I) are closest to the origin and T-R complexes (state IV) are farthest from it.

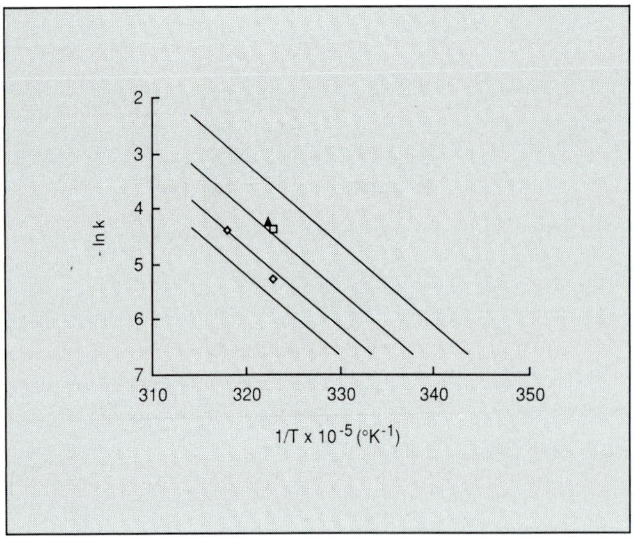

Fig. 3. Arrhenius plot of dissociation constants from patient 4007 with minimal androgen insensitivity. Monolayers labeled at 37 °C with ^{3}H-MB (◇), ^{3}H-DHT (▲) and ^{3}H-MT (□) were chased at various temperatures. Rate constants (table 2) calculated from the slopes of linear dissociation data were plotted against the reciprocal of the absolute temperature. The lines represent the hierarchy of states for A-R complexes in control cells. The lines for control MB-R complexes (state I) are closest to and control T-R complexes (state IV) farthest from the origin.

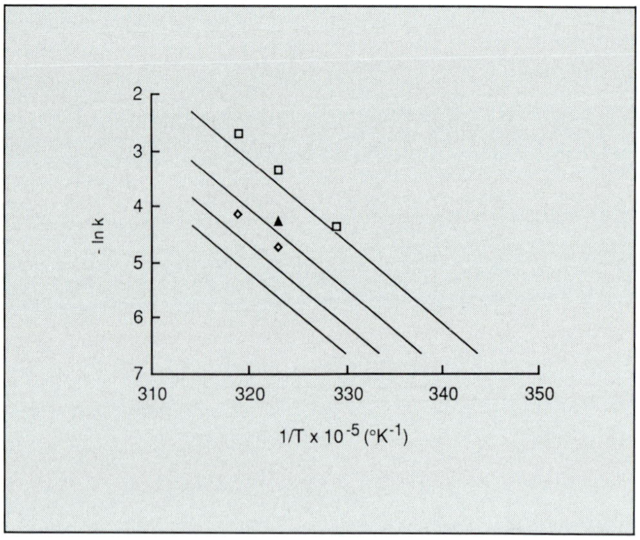

Fig. 4. Arrhenius plot of dissociation constants from patient 3287 with partial androgen insensitivity. Monolayers labeled at 37 °C with ^{3}H-MB (◇), ^{3}-DHT (▲) and ^{3}H-MT (□) were chased at various temperatures. Rate constants (table 2) calculated from the slopes of linear dissociation data were plotted against the reciprocal of the absolute temperature. The lines are defined in the legend to figure 3.

The rate data for both patients are given in table 2. The effect of temperature on the dissociation rate constants of the various A-R complexes within the cells of both patients is shown in figures 3 and 4. In each figure the results are represented by symbols that are superimposed on the hierarchy of energy states for A-R complexes in control cells. Rate constants for the dissociation of MB-R complexes within patient 4007's cells were linear with temperature and when compared to the control hierarchy were in energy state II. They were thus in an atypical energy state one step higher than normal. At 37 °C his DHT-R complexes dissociated abnormally from state III. On the other hand, his MT-R complexes were unaffected and dissociated normally from state III. Patient 3287's various A-R complexes were more severely affected than those of patient 4007. Rate constants for his MT-R complexes were linear with temperature and dissociated from state IV. They were thus in an atypical energy state that was one step higher than normal. At 37 °C the rate constant for his DHT-R complexes indicated dissociation from state III one step higher than normal. Notably, his MB-R complexes dissociated at rates that were between states II and III.

The response of MB- and MT-R complexes in the cells of patients 4007 and 3287 to either thermal stress or prolonged incubation with androgen are shown in figures 5 and 6. In both patients, the complexes were thermolabile; however, they up-regulated normally.

The range, mean and variance of binding capacities (B_{max}) for DHT- and MB-R complexes within control cells, determined under conditions (10 μ*M* cycloheximide) which abolish up-regulation, are presented in figure 7. With MB control cell lines (n = 15) had a mean B_{max} of 36 fmol/mg protein (99% CI, 27.3–44.7); with DHT it was 19.5 (99% CI, 15.7–23.3). The apparent equilibrium affinity constant (K_d) of complexes formed with either androgen was 0.1–0.2 n*M*. The greater mean binding activity with MB than DHT, termed recruitment, was 16.5 fmol/mg protein (99% CI, 12.7–19.9). B_{max} values for patient 4007's DHT- and MB-R complexes were within the range of normals, as were the apparent affinities (K_d = 0.2 n*M*) of these interactions. Recruitment in this patient was also normal. In two separate trials, B_{max} values for patient 3287's MB-R complexes were clearly below the normal range; values for DHT-R complexes, on the other hand, were borderline.

5
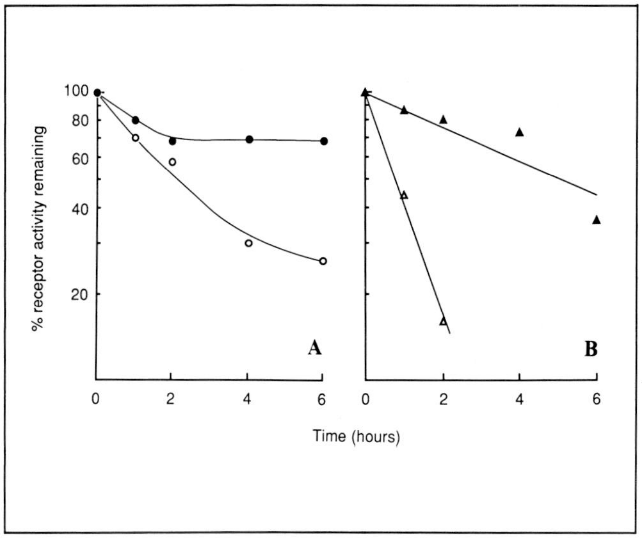

6
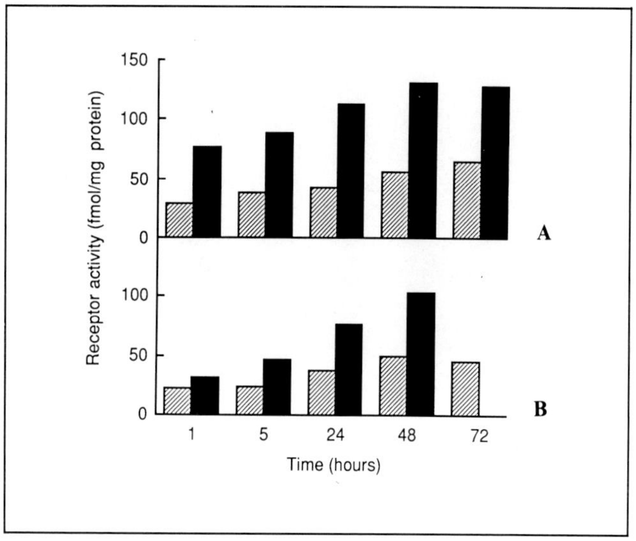

Fig. 5. Thermolability of MB-R complexes in control (●, ▲) and patients' (○, △) GSFs. After incubation for 16 h with 3 n*M* ^{3}H-MB the medium was replaced with one containing 1 n*M* ^{3}H-MB along with 10 μ*M* cycloheximide and the loss of activity in patient 4007's (**A**) and patient 3287's (**B**) cells measured in cell monolayers at 42 °C.

Fig. 6. Up-regulation of A-R complexes in GSFs of patients 4007 (**A**) and 3287 (**B**). Controls (black) and patients' (shaded) GSFs were assayed for androgen receptor activity at various times after incubation with 3 n*M* ^{3}H-MT (**A**) or ^{3}H-MB (**B**).

Fig. 7. Range, mean and variance of binding capacities (B_{max}) for DHT- and MB-R complexes within control (●) GSFs. Replicate monolayers from patients 4007 (■) and 3287 (△, ◇) were incubated with various concentrations (0.05–3.0 n*M*) ^{3}H-DHT or ^{3}H-MB in the presence of 10 μ*M* cycloheximide and B_{max} was determined by Scatchard analysis.

7
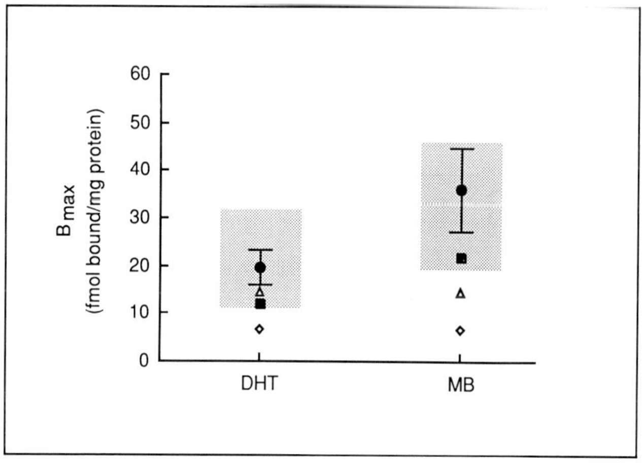

Recruitment of androgen receptor activity in this patient was abnormal even though the apparent affinity of his A-R interactions were normal.

Discussion

Patients with minimal and partial androgen resistance have been previously classified into ligand-sensitive and ligand-independent groups, respectively, according to the intracellular dissociative misbehavior of their various A-R complexes. In the former MT-R complexes dissociate normally, yet both MB- and DHT-R complexes dissociate abnormally fast [1, 9, 10]. A-R complexes from the ligand-independent group dissociate quickly with all 3 ligands [15]. Within the ligand-sensitive group, the mutant androgen receptors are able to

Table 2. Dissociation rate constants for various A-R complexes within mutant GSFs

Strain of patient	Ligand	Temperature °C	Rate constant min^{-1}
4007	MB	37	0.005[1] (n = 3)
		42	0.012 (n = 2)
	DHT	37	0.014
	MT	37	0.013 (n = 2)
3287	MB	37	0.009 (n = 2)
		41	0.016[2] (n = 3)
	DHT	37	0.014[2] (n = 3)
	MT	31	0.013
		37	0.035 (n = 2)
		40	0.067

[1] SD = ± 0.001.
[2] SD = ± 0.003.

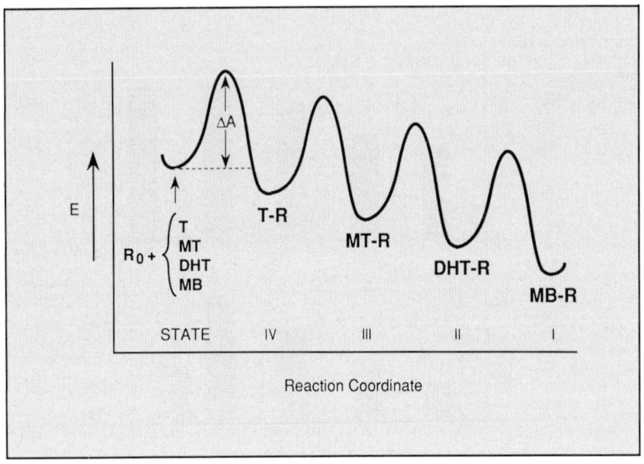

Fig. 8. State transitions of the androgen receptor. This is an energy diagram. The ordinate is in units of relative energy while the abscissa depicts the sequential energy changes occurring in the receptor (R_o) after binding various androgens. The activation energies (ΔA) for transitions between sequential states in the hierarchy need not be the same.

increase (up-regulate) their activity in response to prolonged incubation with MB or MT [9, 10], and to form high-affinity ($K_d < 0.15$ n*M*) complexes with MB or MT [9, 10]; these MB- or MT-R complexes are thermolabile [1] despite the fact that the free mutant receptor is thermostable. The A-R complexes of the ligand-independent group are also thermolabile [1], but they differ from those of the ligand-sensitive group in that they usually have low-affinity K_d values (> 0.3 n*M*) and are unable to up-regulate normally [11].

In the foregoing context, patient 4007 is clearly a member of the ligand-sensitive group. The mutant receptor in patient 3287 is unique: it is similar to those of patients in the ligand-independent group in that its A-R complexes are thermolabile and dissociate at abnormal rates with all 3 ligands, but it can up-regulate normally and form A-R complexes with high-affinity K_d values.

Differences in the dissociation rates of complexes formed between a receptor and various cognate steroids, that bind either as agonists or antagonists, are conventionally assumed to reflect a static steroid-binding domain that has constitutively different 'affinities' for these compounds. According to this assumption, one might expect a mutant androgen receptor to alter its affinity for a set of androgens in a rank order that reflected the affinity of the normal receptor for the same androgens. However, when plotted by the method of Arrhenius, the dissociative behavior of the A-R complexes of both patients departs from this expectation. Instead, their dissociation rates are located in uniquely atypical positions on the hierarchy of dissociative behavior that is expressed by control MB-, DHT-, MT- and T-R complexes. Thus, MB- and DHT-R complexes in the cells of patient 4007 are in energy states II and III, respectively, one state higher than observed for control MB- and DHT-R complexes. On the other hand, his MT-R complexes occupied state III, and thus were unaffected. The more severely affected DHT- and MT-R complexes in the cells of patient 3287 occupied energy states III and IV. They, thus, existed in states that are normally occupied by control MT and T-R complexes, respectively. His MB-R complexes, on the other hand, dissociated at rates between states II and III.

We believe that the patterns generated by these results reflect some essential mechanistic relationship between normal A-R complexes in various states of the hierarchy. A model of this relationship as suggested by these and other data [15] is presented in figure 8. This model views the androgen receptor as a conformationally mobile molecule capable of undergoing a series of *temperature-* and *androgen-dependent* transitions along a common pathway. For example, MB-R complexes are assumed to reach state I by first passing through a T-R complex-like state (IV), a MT-R complex-like state (III) and a DHT-R complex-like state (II). In mutant cells the MB-R complexes are incapable of undergoing some or all of these transitions. Hence, they are 'blocked' in *androgen-inappropriate* states, and display dissociative properties identical to control A-R complexes brought to an *androgen-appropriate* state with a different androgen. Thus, it is apparent that the dissociative properties of an A-R complex in cells are primarily dependent on the energy state of the complex rather than on the particular androgen of which it is composed. According to the hierarchical model, the dissociation rate constants for patient 3287's MB-R complexes, that fall at positions between state II and III, represent intracellular mixtures of complexes in both states.

The comparative apparent equilibrium binding properties (K_d and B_{max} values) of DHT- and MB-R complexes in control and patient cells are also noteworthy because they highlight the complexities of A-R interaction within intact cells. First, the cells of both patients form A-R complexes with high-affinity K_d values, despite the fact that some of these complexes have greatly elevated dissociation rate constants. Such pairs of results have been reported previously by us [1] and others [4, 8]. Patient 3287's MT-R complexes are in energy state IV, like that of normal T-R complexes. Yet, the former have a high-affinity K_d value, whereas the latter have a low-affinity value of K_d [16]. This discordance is underlined by the fact that in the case of other patients [5] various A-R complexes have low-affinity K_d values that are in accord with rapid rates of dissociation typical of A-R complexes in state IV.

Scatchard analysis was originally designed to analyze the kinetics of ligand-receptor interaction in a closed system; that is, when increasing concentrations of a single ligand saturate a receptor that is present at a constant concentration to yield a single species of a ligand-receptor complex, that dissociates at a characteristic rate, and reassociates at the same rate as it was originally formed. These ideal constraints do not apply to the kinetics of A-R complex formation and dissociation within GSFs as evidenced by time-dependent changes in the linearity and slopes of experimental Scatchard plots [5, 10]. Indeed, to explain these changes we have had to postulate a model of A-R interaction in these cells [15] that involves a redistribution of A-R complexes among sequential conformational states, each with its unique set of association and dissociation rate constants.

A second confounding series of observations concerns the comparative B_{max} values for DHT- and MB-R complexes in GSFs of patients and controls. B_{max} values for DHT-R complexes within control and patient 4007's cells, determined under conditions (10 μM cycloheximide) that nullify increases in receptor activity due to upregulation, were significantly lower than replicate values for MB-R complexes. On the other hand, B_{max} values for MB-R complexes in the cells of patient 3287 were below normal, and identical to replicate values determined for DHT-R complexes.

To explain these results, we proffer that kinetic analysis of androgen receptor activity in GSFs is additionally complicated by the fact that these cells contain androgen receptor molecules both in a binding and a nonbinding form. Conversion of nonbinding molecules to the binding form(s) during the binding assay increases androgen receptor activity by a process that is (1) ligand-specific, (2) cycloheximide-resistant, and (3) probably regulated by the concentration of one or another distinct state of A-R complex. According to this hypothesis, MB recruits androgen receptors from the nonbinding pool to a greater degree than does DHT, thereby accounting for the ligand-specific B_{max} differences observed in control cells. Defective recruitment in the GSFs of patient 3287 but not of patient 4007 is thus an additional marker of this patient's androgen receptor aberration. Finally, differential recruitment of androgen receptors must also contribute to the apparent equilibrium binding affinity of A-R complexes in control and mutant cells.

Acknowledgments

This work was supported by the Medical Research Council of Canada Group Grant in Medical Genetics. We are grateful to Nelly Sabbaghian, Rose Lumbroso and Hélène Bourassa for devoted assistance.

References

1 Kaufman M, Pinsky L, Killinger DW: Ligand specific thermal misbehaviour of synthetic androgen-receptor complexes in genital skin fibroblasts of subjects with familial ligand-sensitive androgen resistance. J Steroid Biochem 1986;25:323–331.
2 Griffin JE: Testicular feminization associated with a thermolabile androgen receptor in cultured human fibroblasts. J Clin Invest 1979;64:1624–1631.
3 Kaufman M, Pinsky L, Feder-Hollander R: Defective up-regulation of the androgen-receptor in human androgen insensitivity. Nature 1981;293:735–737.
4 Grino PB, Griffin JE, Cushard WG, et al: A mutation of the androgen receptor associated with partial androgen resistance, familial gynecomastia, and fertility. J Clin Endocrinol Metab 1988;66:754–761.
5 Kaufman M, Pinsky L, Bowin A, et al: Familial external genital ambiguity due to a transformation defect of the androgen receptor that is expressed with dihydrotestosterone and the synthetic androgen methyltrienolone. Am J Med Genet 1984;18:493–507.
6 Brown TR, Maes M, Rothwell SW, et al: Human complete androgen resistance with normal dihydrotestosterone receptor binding capacity in cultured skin fibroblasts: Evidence for a qualitative abnormality of the receptor. J Clin Endocrinol Metab 1982;55:61–69.
7 Pinsky L, Kaufman M, Summitt RL: Congenital androgen insensitivity due to a qualitatively abnormal androgen receptor. Am J Med Genet 1981;10:91–99.
8 Grino PB, Isidro-Gutierrez RF, Griffin JE, et al: Androgen resistance associated with a qualitative abnormality of the androgen receptor and responsive to high dose androgen therapy. J Clin Endocrinol Metab 1989;68:578–584.
9 Pinsky L, Kaufman M, Chudley AE: Reduced affinity of the androgen receptor for dihydrotestosterone but not methyltrienolone in a form of partial androgen resistance. J Clin Invest 1985;75:1291–1296.
10 Pinsky L, Kaufman M, Killinger DK, et al: Human minimal androgen insensitivity with normal dihydrotestosterone-binding capacity in cultured genital skin fibroblasts: Evidence for an androgen-selective qualitative abnormality of the receptor. Am J Hum Genet 1984;36:965–978.
11 Gottlieb B, Kaufman M, Pinsky L, et al: Extracellular correction of the androgen-receptor transformation defect in two families with complete androgen-resistance. J Steroid Biochem 1987;28:279–284.

12 Kaufman M, Pinsky L, Hollander R, et al: Regulation of the androgen receptor by androgen in normal and androgen-resistant genital skin fibroblasts. J Steroid Biochem 1983;18:383–390.
13 Lowry OH, Rosebrough NF, Farr AL, et al: Protein determination with the Folin phenol reagent. J Biol Chem 1951;193:265–275.
14 Scatchard G: The attraction of proteins for small molecules and ions. Ann NY Acad Sci 1949;51:660–672.
15 Kaufman M, Pinsky L: A single-site allosteric model of intracellular androgen-receptor interaction. J Steroid Biochem 1989;32:113–119.
16 Maes M, Sultan C, Zerhouni N, et al: Role of testosterone binding to the androgen receptor in male sexual differentiation of patients with 5α-reductase deficiency. J Steroid Biochem 1979;11:1385–1390.
17 Pinsky L, Kaufman M, Levitsky LL: Partial androgen resistance due to a distinctive qualitative defect of the androgen receptor. Am J Med Genet 1987;27:459–466.
18 Kaufman M, Pinsky L, Simard L, et al: Defective activation of androgen receptor complexes. Mol Cell Endocrinol 1982;25:151–162.
19 Kaufman M, Pinsky L: The dissociation of testosterone and dihydrotestosterone-receptor complexes formed within cultured human genital skin fibroblasts. J Steroid Biochem 1983;18:121–125.

Dr. M. Kaufman
Lady Davis Institute for Medical Research
3755 Côte Ste-Catherine Road
Montréal, Québec H3T 1E2 (Canada)

Discussion

C. Sultan, France: According to what you and other groups have found with molecular biology techniques, do you still think that this step you call activation or transformation of cytosolic receptor is a necessary step for the action of androgen?

M. Kaufman, Canada: There are various states of the androgen receptor and the androgen receptor is brought to these states in normal situations with various different androgens. In abnormal situations the androgen does not bring the receptor to its proper state and therefore the patients that I have described are defective, e.g. they do not undergo normal energy state transitions. Whether this is in fact identical to what other people have classically called 'transformation' remains to be determined but there are androgen receptors in these patients which do not have the same biochemical phenotype as displayed by normal individuals.

C. Sultan: Do you think it is of value to study cytosolic receptors?

M. Kaufman: These studies are all done on intact fibroblasts. These are not cytosolic receptors but total cellular receptors.

Horm Res 1990;33:95–98

0301–0163/90/0334–0095$2.75/0

Structural and Functional Studies of Mammalian Progesterone Receptors

M. Misrahi[a], *H. Loosfelt*[a], *M. Atger*[a], *A. Guiochon-Mantel*[a], *M. Applanat*[a], *A. Bailly*[a], *M.T. Vu Hai-Luu Thi*[a], *P. Lescop*[a], *F. Lorenzo*[a], *P. Bouchard*[b], *E. Milgrom*[c]

[a] Institut National de la Santé et de la Recherche Médicale, Unité 135, Hormones et Reproduction, et
[b] Service d'Endocrinologie, Hôpital de Bicêtre, Le Kremlin-Bicêtre, France

Key Words. Progesterone receptor · Steroid hormone · Gene regulation · Zinc finger protein · Steroid thyroid hormone family

Abstract. During the past years there has been an improvement in our understanding of the molecular mechanism of action of the progesterone receptor (PR). This was due to the obtention of monoclonal antibodies against PR which allowed the first structural analyses and led to the cloning of the genes.

Monoclonal Antibodies against Rabbit and Human Progesterone Receptors

Monoclonal antibodies against the rabbit progesterone receptor (PR) were obtained [1]. They were used to further purify this receptor by immunoaffinity chromatography. Immunoblot studies of the purified PR showed the existence of a major 110-kilodalton (kDa) component and smaller species. The generation of the latter was inhibited by using protease inhibitors [2] during the process of purification, leading to the conclusion that they arose from proteolysis in vivo.

Over 50 monoclonal antibodies were raised against rabbit PR. Antibodies cross-reacting with human PR were used to purify the human PR by immunoaffinity chromatography. Other 30 monoclonal antibodies were produced against the human receptor. Precise mapping of the epitopes in the rabbit [3] and human [4] PRs was performed after cloning of the corresponding genes, using a new method based on immunoprecipitation of truncated proteins obtained by in vitro transcription translation [3].

It could be determined that the lower-molecular-weight components are the same in the rabbit and human, and are due to loss of the N-terminal part of the receptor.

Cloning of Rabbit and Human PRs

Rabbit PR mRNA was first detected by in vitro translation and immunoprecipitation. These experiments confirmed that the PR was synthesized as a 110-kDa form very similar to the native receptor, and thus that no major posttranslational events were involved. Estrogen treatment increased PR through an increase in PR mRNA levels in the rabbit uterus.

The cloning of rabbit PR cDNA was performed in a λgt11 expression library which was screened with polyclonal and monoclonal antibodies. The positive clones were used to isolate clones containing the whole coding region of rabbit PR [5]. The cloning of the human PR was then performed by cross-hybridization using a T47D human breast cancer cell cDNA library screened with rPR cDNA probes [6].

The primary structure of the deduced proteins were compared [7]. The rabbit and human PR have 930 and 933 residues, respectively, and exhibit an 88.3% overall homology. Moreover, this homology is especially focused in regions which must be important for receptor function.

The central cysteine-rich DNA-binding region is 100% conserved between both receptors. By analogy with other transcriptional factors it could display a bidigital tridimensional arrangement stabilized by zinc. The COOH hydrophobic hormone-binding domain is 99.5% homologous in the two mammalian receptor types. The N terminus is the most divergent between the rabbit and human PR and also between different receptors of the steroid thyroid hormone family [8]. In PRs it is very rich in proline and acidic. The difference in composition and codon usage of the N-terminal part of PRs has led to the hypothesis that it could have been added secondarily to the protein by exon shuffling [7].

Northern blot analysis using PR cDNAs showed the existence of a doublet in rabbit uterus (~5,900 and 6,600 nucleotides) [5] and several species (~11,000, 5,900, 5,100, 4,300, 3,700 and 2,900 nucleotides) in the human uterus T47D and MCF7 cell lines [6].

This heterogeneity is not due to variations in the coding regions: similar patterns were observed in Northern blots using short probes corresponding to the N- and C-terminal ends of the PR, but rather to variations in the 3′ noncoding regions. Three polyadenylation sites were found in rabbit PR mRNA (3,058, 3,300 and 3,553 nucleotides after the TGA stop codon) [9].

In vitro transcription translation experiments yielded a major 110-kDa protein. The initiation ATG is thus the only main codon which is used to start the transduction in vitro. These experiments are in agreement with previous data suggesting that the A form is derived from the B form by artifactual proteolysis [10].

Mutagenesis Experiments of Rabbit PR

The mechanism of nuclear localization of the rabbit PR was studied by mutagenesis experiments. Two main mechanisms were identified [11].

The first one involves the existence of nuclear targeting sequences: the main one is localized around amino acids 638–643 and is constitutively active. This sequence is very basic and exhibits similarity to the SV40 large T antigen nuclear signal sequence.

Another mechanism consists in the activation of the DNA-binding domain either through the binding of the hormone or by deletion of the steroid-binding region (constitutive receptor).

Direct interactions between monomers of the receptors were shown using co-transfection experiments with receptors having deletions of epitopes recognized by monoclonal antibodies. These interactions occurred throughout the steroid-binding domains of both monomers.

Chromosome Mapping of Human PR Gene

The human PR gene was localized to chromosome 11q22–q23 using two different probes corresponding to the 5′ and 3′ part of the cDNA [12]. This localization has been confirmed by Mattei et al. [13] and differs from the previous report of Law et al. [14].

Promoter and 5′ Flanking Region of Rabbit PR Gene

The PR is submitted to strict hormonal regulations: it is induced by estrogen and down-regulated by progesterone. These regulations seem to occur at the mRNA level. The tissue-specific expression of the PR gene is also precisely defined. To initiate the study of these regulatory mechanisms, 5′ genomic fragments of the rabbit PR gene were cloned and sequenced [9]. The 5′ noncoding region extends 699–712 basepairs (bp) and contains two transcription start sites. The rPR promoter region lacks a typical TATA box. A CAACT sequence is present at position –100 and one consensus site for the transcription factor Sp1 at position –51. Several sequences resembling the consensus for estrogen and progesterone response elements were observed in the 5′ flanking region of the gene and thus argue in favor of a transcriptional regulation of the PR gene by both steroids.

Some of these consensus sequences corresond to effective receptor binding sites as shown by deoxyribonuclease (DNaseI) footprinting experiments with purified PR and estrogen receptor (ER) [A. Bailly, unpubl. data]. Functional analysis by in vivo transfection experiments are currently in progress.

Study of Interaction of PR with DNA

Steroid hormone receptors modulate the expression of specific genes after binding to specific sequences, hormone-responsive elements, located most often in the upstream region of target genes [15]. Using DNaseI footprinting experiments, the sequences recognized by purified PR on the uteroglobin gene were precisely determined [16]. They are localized as far as –2,650 bp upstream of the transcription initiation. The same results were obtained with receptor activated by ligand binding, or independently of the phosphorylation status of the receptor. However, in vivo hormone is indispensable for acquisition of tight nuclear binding of the receptor. This supports the hypothesis of the existence in vivo of an inhibitor stabilizing the receptor in a 'nonactivated state' [16]. The action of the antiprogestin RU38486 was studied in vitro and in vivo. Purified RU38486-PR complexes recognized the same target sequences on the progesterone-responsive uteroglobin gene. Moreover, these experiments were confirmed by a functional test. Taking advantage of a constitutively active PR, obtained by deletion of amino acids 663–930 of the steroid-binding domain of rabbit PR, competition experiments were performed with wild-type receptor complexed to R5020 or RU38486. This showed that the receptor liganded with the agonist or the antagonist is activated, able to bind DNA, and that both complexes compete for the same hormone-responsive elements in vivo [17]. The action of the antagonist must therefore occur at further steps of hormone action.

Phosphorylation of the PR

The PR undergoes two phosphorylation steps: a basal phosphorylation and a hormone-dependent phosphorylation. The latter provokes a characteristic upshift of receptor mobility in sodium dodecyl sulfate-polyacrylamide gels [18]. A kinase has been identified which copurifies with the receptor but it is not hormone-regulated in vitro [19].

The role of the hormone-dependent phosphorylation is not known. It is not necessary for the ability of the receptor to bind to specific hormone-responsive elements in vitro. It could be involved in further steps of hormone action, leading to modulation of transcriptional activity of target genes. Alternatively, it could display a role in receptor down-regulation.

Immunocytochemical Studies of Mammalian PR

Monoclonal antibodies against rPR were used to study the cellular and subcellular distribution of the receptor by light microscopy in various tissues of the rabbit and guinea pig [20]. The receptor was found to be localized in the nucleus in the presence or in the absence of hormone. However, the nonliganded receptor does not exhibit tight nuclear binding. This explains why it was easily extracted from the nucleus by biochemical methods and found in the 'cytosol'.

The subcellular distribution of PR was further studied at the electron-microscopic level in uterine stromal cells using an immunogold method [21]. In the absence of progestin most of PR was localized in the nucleus where it was randomly associated with condensed chromatin. A very small staining was observed to be associated with the rough endoplasmic reticulum. This could reflect the newly synthesized receptor or indicate a post-transcriptional function of the receptor. After hormone administration the chromatin underwent extensive rearrangement while the receptor was displaced toward the borders of condensed chromatin, both regions being the most active sites of gene transcription.

Clinical Applications of Immunological Studies of PR

Determination of steroid hormone receptors has become widely used in the management of breast cancers. The immunocytochemical detection of PR by the peroxidase-antiperoxidase method was used to study human breast cancers [22]. There was a good correlation with the concentration of PRs measured by steroid-binding assay in tumor extracts. Heterogeneity of staining was observed, suggesting heterogeneity in the PR content of breast cancer cells.

The immunocytochemical detection of human PR displays several advantages over the steroid-binding method. It can detect the free hormone receptors and the tightly nucleus-bound steroid receptor complexes. Moreover, this method could give information on the cellular distribution of receptors within a tumor.

ER and PR were studied on endometrial fragments obtained by endometrial aspiration of endometrial biopsy throughout the menstrual cycle [23]. Important changes in the proportion of immunolabeled cells and in the intensity of labeling were observed. Three characteristic patterns could be defined. In particular, there was an intense ER and PR staining during the midcycle period correlated with the increased estradiol secretion at this time. There was a decrease and even a disappearance in glandular cells of ER and PR during the late luteal phase under the cumulative effect of progesterone during this period of the cycle. The level of these steroid receptors during the menstrual cycle could, therefore, be used as a functional assay which may be useful in studies of infertility in women.

References

1 Logeat F, Vu Hai-Luu Thi MT, Fournier A, et al: Monoclonal antibodies to rabbit progesterone receptor: Cross-reaction with other mammalian progesterone receptors. Proc Natl Acad Sci USA 1983;80:6456–6459.

2 Loosfelt H, Logeat F, Vu Hai-Luu Thi MT, Milgrom E: The rabbit progesterone receptor: Evidence for a single steroid-binding subunit and characterization of receptor mRNA. J Biol Chem 1984;259:14196–14202.
3 Lorenzo F, Jolivet A, Loosfelt H, Vu Hai-Luu Thi MT, et al: A rapid method of epitope mapping: Application to the study of immunogenic domains and to the characterization of various forms of rabbit progesterone receptor. Eur J Biochem 1988;176: 53–60.
4 Vu Hai-Luu Thi MT, Jolivet A, Ravet V, et al: Novel monoclonal antibodies against human uterine progesterone receptor: Mapping of receptor immunogenic domains. Biochem J 1989; 260:371–376.
5 Loosfelt H, Atger M, Misrahi M, et al: Cloning and sequence analysis of rabbit progesterone-receptor complementary DNA. Proc Natl Acad Sci USA 1986;83:9045–9049.
6 Misrahi M, Atger M, d'Auriol L, et al: Complete amino acid sequence of the human progesterone receptor deduced from cloned cDNA. Biochem Biophys Res Commun 1987;143:740–748.
7 Misrahi M, Atger M, d'Auriol L, et al: Human and rabbit progesterone receptors: Comparison of amino acid sequences deduced from cloned cNDAs; Steroid Hormone Action. New York, Alan R. Liss, 1988, pp 127–139.
8 Evans RM: The steroid and thyroid hormone receptor superfamily. Science 1988;240:889–895.
9 Misrahi M, Loosfelt H, Atger M, et al.: Organization of the entire rabbit progesterone receptor mRNA and of the promoter and 5′ flanking region of the gene. Nucleic Acids Res 1988;16: 5459–5472.
10 Logeat F, Pamphile R, Loosfelt H, et al: One-step immunoaffinity purification of active progesterone receptor: Further evidence in favor of the existence of a single steroid binding subunit. Biochemistry 1985;24:1029–1035.
11 Guiochon-Mantel A, Loosfelt H, Lescop P, et al: Mechanisms of nuclear localization of the progesterone receptor: Evidence for interaction between monomers. Cell 1989;57:1147–1154.
12 Rousseau-Merk MF, Misrahi M, Loosfelt H, et al: Localization of the human progesterone receptor gene to chromosome 11q22–11q23. Human Genet 1987;77:280–282.
13 Mattei MG, Krust A, Stropp U, et al: Assignment of the human progesterone receptor to the q22 band of chromosome. Hum Genet 1988;78:96–97.
14 Law ML, Kao FT, Wei Q, et al: The progesterone receptor gene maps to human chromosome band 11q13, the site of the mammary oncogene int-2. Proc Natl Acad Sci USA 1987;84:2877–2881.
15 Beato M: Gene regulation by steroid hormones. Cell 1989;56: 335–344.
16 Bailly A, Le Page C, Rauch M, Milgrom E: Sequence-specific DNA binding of the progesterone receptor to the uteroglobin gene: Effects of hormone, antihormone and receptor phosphorylation. EMBO J 1986;5:3235–3241.
17 Guiochon-Mantel A, Loosfelt H, Ragot T, et al: Receptors bound to antiprogestin form abortive complexes with hormone responsive elements. Nature 1988;336:695–698.
18 Logeat F, Le Cunff M, Pamphile R, Milgrom E: The nuclear-bound form of the progesterone receptor is generated through a hormone dependent phosphorylation. Biochem Biophys Res Commun 1985;131:421–427.
19 Logeat F, Le Cunff M, Rauch M, et al: Characterization of a casein kinase which interacts with the rabbit progesterone receptor: Differences with the in vivo hormone-dependent phosphorylation. Eur J Biochem 1987;170:51–57.
20 Perrot-Applanat M, Logeat F, Groyer-Picard MT, Milgrom E: Immunocytochemical study of mammalian progesterone receptor using monoclonal antibodies. Endocrinology 1985;116: 1473–1484.
21 Perrot-Applanat M, Groyer-Picard MT, Logeat F, Milgrom E: Ultrastructural localization of the progesterone receptor by an immunogold method: Effect of hormone administration. J Cell Biol 1986;102:1191–1199.
22 Perrot-Applanat M, Groyer-Picard MT, Lorenzo F, et al: Immunocytochemical study with monoclonal antibodies to progesterone receptor in human breast tumors. Cancer Res 1987;47: 2652–2661.
23 Garcia E, Bouchard P, De Brux J, et al: Use of immunocytochemistry of progesterone and estrogen receptors for endometrial dating. J Clin Endocrinol Metab 1988;67:80–87.

M. Misrahi, MD
Institut National de la Santé et de la Recherche Médicale
Unité 135
Hormones et Reproduction
Hôpital de Bicêtre
F–94270 Le Kremlin-Bicêtre (France)

Horm Res 1990;33:99–104

0301–0163/90/0334–0099$2.75/0

Regulation of Glucose Transporters in Diabetes

E. Karnieli, M. Armoni

Metabolic Unit, Endocrine Institute, Rambam Medical Center, and Faculty of Medicine, Technion-Israel Institute of Technology, Haifa, Israel

Key Words. Glucose transport · Glucose transporter, intrinsic activity · Translocation · NIDDM · Streptozotocin-diabetic rats · Insulin therapy

Abstract. It is now widely accepted that insulin stimulates glucose metabolism in its target tissues via recruitment of transporters from a large intracellular pool to the plasma membrane. Recent studies, however, suggest a two-step model for insulin action, of transporter translocation and transporter activation. Data confirming this hypothesis for the first time are presented. It is shown that insulin significantly enhances the intrinsic activity of glucose transporters in human and rat adipose cells, in physiological as well as in diabetic state. The functional activity of transporters is impaired in the diabetic state, but surprisingly, 'diabetic' transporters exhibit normal or even enhanced intrinsic activity. In both noninsulin-dependent diabetes mellitus and streptozotocin-diabetic rats, insulin resistance is associated with 50% transporter depletion in the intracellular pool, thus leading to a decreased number of transporters appearing in the plasma membrane in response to insulin. It is concluded that impaired glucose transport in diabetes is secondary (1) to intracellular transporter depletion, and (2) to the presence of inhibitory factors interfering with the full expression of glucose transporters at the plasma membrane, thus contributing to postreceptor insulin resistance.

Introduction

Binding of insulin to its receptor initiates a series of biochemical events leading to a variety of cellular actions. We will hereby review some of the mechanisms underlying insulin's ability to stimulate glucose transport in adipose and muscle tissues, and examine the contribution of impairments of the glucose transporters to the postreceptor insulin resistance in diabetes.

A fundamental action of insulin is its ability to stimulate glucose metabolism in target tissues. The first step in the intracellular metabolism of glucose is its entry across the plasma membrane, a process that is under stringent regulation in adipose and muscle cells. The extreme sensitivity of fat cells to insulin, and the accessibility of viable tissue, made it the tissue of choice as early as the 1960s to examine the effects of insulin on cellular glucose transport [1–5]. In adipocytes, insulin acts very rapidly to stimulate glucose transport, even though there is a lag period of up to 45 s at 37 °C [6, 7]. Exposing intact cells to maximally stimulating hormone concentrations results in a 10- to 30-fold increase in glucose transport rate within 10–15 min, with a $t_{1/2}$ of 2.5–4 min at 37 °C [8, 9]. Insulin stimulation of cellular glucose transport can occur via an enhancement of the transporter affinity (K_m), and/or its maximal capacity (V_{max}), towards its substrate. Increased V_{max} can result from increased number and/or functional activity of transporters at the cell surface. Kinetic analyses using glucose analogues have uniformly shown that insulin stimulates the glucose transport rate by enhancing V_{max} [10–13], while results regarding K_m are still controversial [14–16]. Insulin-induced increase in V_{max} can only partially be accounted for by the increased number of transporters recruited to the plasma membrane [11]. Latest works in both physiological [17–19] and pathophysiological states [20–22] strongly suggest a two-step model of transporter translocation to, and activation at, the plasma membrane.

The Translocation Theory

Independent studies from Cushman's laboratory [8, 11, 23] and Suzuki and Kono's laboratory [24–26] concurrently led to the discovery of an intracellular pool of glucose transporters in rat adipocytes, which was associated with a low-density microsomal membrane fraction enriched with enzymes of the Golgi apparatus [24, 27]. Cushman's group [8, 11, 23] applied the *D*-glucose-inhibitable cytochalasin-B-binding assay to quantify the number of transporters in subcellular membrane fractions prepared from isolated rat adipocytes, as detailed previously [23]. Basal transporter levels in plasma membranes were as low as 5 pmol/mg membrane protein, while in the low-density microsomes 16-fold higher levels were found. Upon insulin stimulation, a dramatic redistribution of transporters was observed. Insulin induced a 5-fold increase in plasma-membrane-associated transporters, while concomitantly decreasing their level in the intracellular pool by about 60% [8, 24]. Kono's group [24–26], using a different technique of subcellular fractionation on linear sucrose gradient, followed by measurements of *D*-glucose transport in the subfractions after their reconstitution to liposomes, showed similar insulin-induced redistribution of transporters. Both groups concluded that insulin stimulates glucose transport by inducing the translocation of glucose transporters from a large intracellular pool to the plasma membrane. This process was further shown to be rapid ($t_{1/2}$ = 2.5 min), reversible, hormone-concentration-dependent [8], energy- and temperature-dependent, but protein-synthesis-independent [25]. The translocation model has now been validated in different insulin-responsive tissues such as rat diaphragm [28] and cardiac muscle [29]. Recently, we have demonstrated the existence of a similar insulin-induced translocation process in human adipose cells [30], that is subjected to sex differences [19] and site differences [31].

Further validation of the translocation theory came from studies in adipocytes using antibodies against the purified human erythrocyte glucose transporter [32]. In these studies, plasma and low-density microsomal membrane fractions prepared from basal and insulin-stimulated adipocytes were incubated with antibodies raised against the purified human erythrocyte glucose transporter on Western blots. A 45- to 50-kilodalton protein cross-reacted with the antiserum, and the relative amounts of this protein in the membrane fractions were in approximate agreement with the number of transporters determined by the cytochalasin-B-binding assay. However, the poor cross-reactivity obtained in these experiments suggested the existence of another transporter species in target tissues for insulin, that is distinct from the human erythrocyte type, as will be discussed below.

The translocation theory is now largely accepted as a model for insulin's stimulatory effect on glucose transport. According to this model, binding of insulin to its receptor generates signals, the identity of which is still debatable. This results in a rapid exocytotic movement of glucose transporters from an intracellular pool to the plasma membrane, followed by binding and fusion with it. The steps of transporter translocation to and fusion with the plasma membrane are now under intensive investigation in several laboratories, since they comprise a crucial regulatory stage in which the transporter is subjected to additional activation, leading to the enhancement of its intrinsic activity. Derangements of this activation process in altered metabolic states like diabetes or fasting may play a role in the development of insulin resistance at the transporter level.

The Insulin-Responsive Glucose Transporter

Recent studies on the primary sequences of membrane transport proteins with similar function but from different evolutionary origin have revealed that facilitated glucose transport in mammalian cells is mediated by a family of structurally related proteins, and cDNA clones encoding different members of this family were identified. Complementary DNA clones obtained from the human hepatoma (HepG2) cell line [33] and from rat brain [34] were shown to have over 97% identity in the deduced amino acid sequence. The HepG2/rat brain sequence probably also encodes the human erythrocyte glucose transporter, but is unrelated to either one of two additional cDNAs encoding (1) the adult liver and β-cell transporters [35], and (2) the Na^+-linked transport protein of the intestine and kidney cells [36].

Presently, a large body of evidence has accumulated suggesting that none of the existing transporter proteins has the predicted features of a putative 'insulin-responsive' carrier. The existence of such a distinct insulin-regulatable molecule was first suggested by the poor cross-reactivity between antibodies for either the HepG2 or rat brain transporters, and the rat adipose glucose transporter [37, 38]. More direct evidence was obtained using the specific monoclonal antibody (mAb) 1F8 that immunolabeled a novel transporter protein in insulin target tissues and underwent insulin-induced translocation to a much greater extent than the HepG2/rat brain transporter [39]. Additional clues came from experiments showing a consistent discrepancy between insulin effect to stimulate glucose transport and the extent of transporter translocation, as assayed by the cytochalasin-B-binding activity [19–22, 30, 31]. This discrepancy can be explained by the existence in adipose cells of a second pool of insulin-regulatable transporters, as recognized by the specific mAb 1F8. In a search for such yet new molecule, James et al. [40] and Birnbaum [41] have cloned and sequenced cDNAs isolated from rat adipose and muscle cells, which encode a protein that is 65% homologous to the HepG2/rat brain sequence and is expressed exclusively in insulin-responsive tissues.

These findings, however, do not exclude the possibility of a secondary requirement for individual activation by insulin of the translocated transporters, as suggested by many investigators [16–22]. And indeed, we have recently acquired data confirming this hypothesis in both human and rat adipocytes, in physiological as well as in the diabetic state. Simultaneously measuring cytochalasin-B-binding and *D*-glucose transport activities in purified subcellular fractions reconstituted into liposomes, we have shown, directly for the first time, that 30 min preincubation of adipocytes with 7 n*M* insulin significantly enhances the intrinsic activity of the glucose transporter [in preparation]. Therefore, the discrepancy between insulin fold stimulation of glucose transport and transporter translocation can be explained by a two-step model for insulin action, of carrier translocation and a secondary step of enhancing the transporter intrinsic activity during and/or after its translocation to the plasma membrane. Thus, the muscle/adipose transporter can be regarded as insulin-regulatable not only by being translocable, but also intrinsically activated by insulin. We have found this activation to be protein-dependent [42], however, the exact nature of the process is still under investigation.

Insulin Resistance in Diabetes

The discovery of a unique insulin-responsive transporter bare strong clinical importance, since a gene encoding transporter only in insulin target tissues would be an obvious target for a genetic defect causing predisposition towards insulin-resistant states such as noninsulin-dependent diabetes mellitus (NIDDM) and insulin-dependent diabetes mellitus (IDDM).

Insulin Resistance in NIDDM

In NIDDM, peripheral insulin resistance is a major metabolic abnormality contributing to hyperglycemia. In recent years, studies in NIDDM and obese patients have shown that a defect distal to receptor binding in the insulin action cascade is a major abnormality underlying insulin's reduced ability to stimulate glucose uptake [43–47]. In adipocytes isolated from these patients, decreased basal and insulin-stimulated glucose transport activity found in vitro correlated well with in vivo expression of the postreceptor defect. We [48] and Garvey et al. [49] found that NIDDM is associated with decreased levels of transporters in the low-density microsomes, compared to matched normoglycemic controls. This intracellular transporter depletion led to a reduction in the amount of transporters recruited to the plasma membrane, resulting in decreased stimulation of transport rates in response to insulin. However, in both studies, a discrepancy was observed, between a major decrease in insulin stimulation of glucose transport and the extent of impairment in translocation. This could be attributed, at least partially, to a 50% attentuated insulin effect to enhance the functional activity of outer cell transporters. As the transporter function depends on its intrinsic activity, we went one step further to directly assess this activity of the 'diabetic' transporter, compared to controls. Surprisingly, transporters from NIDDM and control adipocytes had similar intrinsic activities (except for a 30% depletion in the diabetic basal intracellular pool). Insulin significantly increased these values in both NIDDM and controls by 20–40%. The disparity between impaired functional activity of the transporter in lieu of intact intrinsic activity suggests that, beyond intracellular transporter depletion, impaired glucose transport in NIDDM can result from inhibitory factors interfering with the full expression of the transporter at the plasma membrane, thus contributing to postreceptor insulin resistance.

Insulin Resistance in Streptozotocin-Induced Diabetic Rats

We have also studied the streptozotocin-induced diabetic rat, as a model for IDDM [16, 20, 50]. In this model, insulin resistance was associated with about 50% transporter depletion in the intracellular pool, leading to a decreased number of transporters appearing in the plasma membrane in response to insulin. The functional activity of the outer cell transporters was impaired in untreated diabetes and significantly improved upon 8-day insulin therapy [51, 52]. Surprisingly, however, when we examined the transporter intrinsic activity, not only was it undamaged, but was 3- to 6-fold higher in diabetes than in control rats. Thus, like in NIDDM, it is suggestive that postreceptor insulin resistance is associated with the presence of inhibitory factors that suppress the full expression of transporters at the plasma membrane.

Impact of Insulin Therapy

A convergence has been shown between in vitro [53, 54] and in vivo [55, 56] findings regarding insulin treatment of NIDDM. We [16, 20] have examined the effects of insulin therapy on the reversal of insulin resistance in the streptozotocin-induced diabetic rat model. Our data show that, while insulin therapy for 8 days restored the cellular content of transporters to control levels, it led to a supranormal 3-fold enhancement of cellular glucose transport rates. Thus, it appears that the functional activity of the transporters, which is impaired in the diabetic state, is overenhanced upon insulin therapy. Lately, Garvey et al. [57] have presented data indicating that both the major transporter depletion in diabetes and its restoration upon 8-day insulin therapy are due to changes in the expression of the insulin-responsive transporter molecule, while the HepG2 species is unaffected.

Conclusion

Insulin regulates glucose transport in its target tissues in a two-step mechanism, namely the translocation of transporters from an intracellular pool to, and enhancement of, their intrinsic activity at the plasma membrane. The major reduction in glucose transport in diabetes may be secondary to transporter depletion in the intracellular pool, and to inhibitory factors intefering with the full expression of transporters at the plasma membrane level in the intact cell. Reversal of diabetes by insulin therapy is associated with restored levels of glucose transporter and a transient protein-synthesis-dependent enhancement of the functional activity. In insulin-treated and untreated diabetic rat models, transporter regulation is exerted mainly on the insulin-regulatable species. Likewise, in NIDDM, insulin resistance and its reversal could be mediated via changes in the amount of muscle glucose transporter mRNA and protein expression in the cell. The nature of the regulatory factors involved in this process is now being studied.

References

1 Winegard AI, Renold AE: Studies on rat adipose tissue in vitro. J Biol Chem 1958;233:267–272.
2 Crofford OB, Renold AE: Glucose uptake by incubated rat epididymal adipose tissue: Rate limiting steps and site of insulin action. J Biol Chem 1965;240:14–21.
3 Crofford OB, Renold AE: Glucose uptake by incubated rat epididymal adipose tissue: Characteristics of the glucose transport system and action of insulin. J Biol Chem 1965;240:3237–3244.
4 Levine R, Goldstein M: On the mechanism of action of insulin. Recent Prog Horm Res 1955;11:343-380.
5 Park CR, Reinwein D, Henderson MJ, et al: The action of insulin on the transport of glucose through the cell membrane. Am J Med 1959;26:674–684.
6 Haring HU, Kemmler W, Renner R, Hepp KD: Initial lag phase in the action of insulin on glucose transport and cAMP levels in fat cells. FEBS Lett 1978;95:177–180.
7 Ciaraldi TP, Olefsky JM: Coupling of insulin receptors to glucose transport: A temperature dependent time lag in activation of glucose transport. Arch Biochem Biophys 1979;193:221–231.
8 Karnieli E, Zarnowski MJ, Hissin PJ, et al: Insulin-stimulated translocation of glucose transport systems in the isolated rat adipose cell: Time course, reversal, insulin concentration dependency and relationship to glucose transport activity. J Biol Chem 1981;256:4772–4777.
9 Ciaraldi TP, Olefsky JM: Relationship between deactivation of insulin stimulated glucose transport and insulin dissociation in isolated rat adipocytes. J Biol Chem 1980;255:327–330.
10 Vinten J, Gliemann J, Osterlind K: Exchange of 3-O-methylglucose in isolated fat cells: Concentration dependence and effect of insulin. J Biol Chem 1976;251:794–800.
11 Wardzala LJ, Cushman SW, Salans LB: Mechanism of insulin action on glucose transport in the isolated rat adipose cell: Enhancement of the number of functional transport systems. J Biol Chem 1978;253:8002–8005.
12 Olefsky JM: Mechanisms of the ability of insulin to activate the glucose transport system in rat adipocytes. Biochem J 1979;172: 137–145.
13 Whitesell RR, Gliemann J: Kinetic parameters of transport of 3-O-methylglucose and glucose in adipocytes. J Biol Chem 1979; 254:5276–5283.
14 Whitesell RR, Abumrad NA: Increased affinity predominates in insulin stimulation of glucose transport in the adipocytes. J Biol Chem 1985;260:2894–2899.
15 Toyoda N, Flanagan JE, Kono T: Reassessment of insulin effects on the V_{max} and K_m values of hexose transport in isolated rat epididymal adipocytes. J Biol Chem 1987;262:2737–2745.
16 Rafaeloff R, Armoni M, Karnieli E: Insulin therapy modulates the affinity of glucose transporters in adipocytes isolated from

diabetic rats (abstract 316). 46th Annu Meet ADA, Anaheim, June 1986. Diabetes 1986;35(suppl 1):79a.
17 Calderhead DM, Lienhard GE: Labeling of glucose transporters at the cell surface in 3T3-L1 adipocytes. J Biol Chem 1988;263: 12171–12174.
18 Obermaier-Kusser B, Muhlbacher C, Mushack J, et al: Further evidence for a two-step model of glucose transport regulation. Biochem J 1989;261:699–705.
19 Armoni M, Rafaeloff R, Barzilai A, et al: Sex differences in insulin action on glucose transport and transporters in human omental adipocytes. J Clin Endocrinol Metab 1987;65:1141–1146.
20 Karnieli E, Armoni M, Cohen P, et al: Reversal of insulin resistance of diabetic rats by insulin therapy: Restoration of the pool of glucose transporters and enhancement of glucose transport activity. Diabetes 1987;36:925–931.
21 Kahn BB, Cushman SW: Mechanism of markedly hyperresponsive insulin stimulated glucose transport activity in adipose cells from insulin-treated streptozotocin diabetic rats: Evidence for increased glucose transport intrinsic activity. J Biol Chem 1987; 262:5118–5124.
22 Kahn BB, Simpson IA, Cushman SW: Divergent mechanisms for the insulin resistant hyperresponsive glucose transport in adipose cells from fasted and refed rats. J Clin Invest 1988;82: 691–699.
23 Cushman SW, Wardzala LJ: Potential mechanism of insulin action on glucose transport in the isolated rat adipose cell. J Biol Chem 1980;255:4758–4762.
24 Suzuki K, Kono T: Evidence that insulin causes translocation of glucose transport activity to the plasma membrane from an intracellular storage site. Proc Natl Acad Sci USA 1980;77: 2542–2545.
25 Kono T, Suzuki K, Dansey LE, et al: Energy dependent and protein synthesis-independent recycling of the insulin sensitive glucose transport mechanism in fat cells. J Biol Chem 1981;256: 6400–6407.
26 Kono T, Robinson FW, Blevins TI, Ezaki O: Evidence that translocation of glucose transport activity is the major mechanism of insulin action on glucose transport in fat cells. J Biol Chem 1982;257:10942–10947.
27 Simpson IA, Yver D, Hissin PJ, et al: Insulin-stimulated translocation of glucose transporters in the isolated rat adipose cell: Characterization of subcellular fractions. Biochim Biophys Acta 1983;763:393–407.
28 Wardzala LJ, Jeanrenaud B: Identification of *D*-glucose inhibitable cytochalasin B binding site as the glucose transporter in rat diaphragm plasma and microsomal membranes. Biochim Biophys Acta 1981;730:49–52.
29 Watanabe T, Smith MM, Robinson FW, Kono T: Insulin action on glucose transport in cardiac muscle. J Biol Chem 1984;259: 13117–13122.
30 Karnieli E, Rafaeloff R, Barzilai A, Armoni M: Distribution of glucose transporters in membrane fractions isolated from human adipose cells: Relation to cell size. J Clin Invest 1986;78: 1051–1055.
31 Karnieli E, Moscona R, Rafaeloff R, et al: Discrepancy between glucose transport and transporters in human femoral adipocytes. Am J Physiol 1989;256:E179–E185.
32 Wheeler TJ, Simpson IA, Sogin DC, et al: Detection of the rat adipocyte glucose transporter with an antibody against the human red blood cell glucose transporter. Biochem Biophys Res Commun 1982;105:89–95.
33 Mueckler M, Caruso C, Baldwin SA, et al: Sequence and structure of the human glucose transporter. Science 1985;229:941–945.
34 Birnbaum NJ, Haspel HC, Rosen OM: Cloning and characterization of cDNA encoding the rat brain glucose transporter protein. Proc Natl Acad Sci USA 1986;83:5784–5788.
35 Thorens B, Sarkar HK, Kaback HR, Lodish HF: Cloning and functional expression in bacteria of a novel glucose transporter present in liver, intestine, kidney and β-pancreatic islet cells. Cell 1988;55:281–290.
36 Hediger MA, Coady MJ, Ikeda ST, Wright EM: Expression cloning and cDNA sequencing of Na^+/glucose cotransporter. Nature 1987;330:379–381.
37 Wang C: The *D*-glucose transporter is tissue specific: Skeletal muscle and adipose tissue have a unique form of glucose transporter. J Biol Chem 1987;262:15689–15695.
38 Oka Y, Asano T, Shibasaki Y, et al: Studies with antipeptide antibody suggest the presence of at least two types of glucose transporter in rat brain and adipocyte. J Biol Chem 1988;263: 13432–13439.
39 James DE, Brown R, Navarro J, Pilch PF: Insulin-regulatable tissue express a unique insulin-sensitive glucose transport protein. Nature 1988;333:183–185.
40 James DE, Strube M, Mueckler M: Molecular cloning and characterization of an insulin-regulatable glucose transporter. Nature 1989;338:83–87.
41 Birnbaum NJ: Identification of a novel gene encoding an insulin-responsive glucose transporter. Cell 1989;57:305–315.
42 Karnieli E, Hueckst eadt TP: Cycloheximide modulates the intrinsic activity of the glucose transporter (abstract 120). 48th Annu Meet ADA, New Orleans, June 1988. Diabetes 1988; 37(suppl 1):30a.
43 Reaven GM, Bernstein-Davis B, Olefsky JM: Non-ketotic diabetes mellitus: Insulin deficiency or insulin resistance? Am J Med 1976;60:80–88.
44 Kolterman OG, Gray RS, Griffin J, et al: Receptor and postreceptor defects contribute to the insulin resistance in non-insulin dependent diabetes mellitus. J Clin Invest 1981;68:957–969.
45 Rizza RA, Mandarino LJ, Gerich GE: Mechanisms of insulin resistance in man: Assessment using the insulin dose response curve in conjunction with insulin receptor binding. Am J Med 1981;70:169–176.
46 Ciaraldi TP, Kolterman OG, Scarlett JA, et al: Role of the glucose transport system in the postreceptor defect on non-insulin dependent diabetes mellitus. Diabetes 1982;31:1016–1022.
47 Kashiwagi A, Verso MA, Andrews J, et al: In vitro insulin resistance of human adipocytes isolated from subjects with non-insulin-dependent diabetes mellitus. J Clin Invest 1983;72:1246–1254.
48 Armoni M, Harel C, Eitan A, et al: Mechanism of insulin resistance in NIDDM patients: Decrease of adipocyte glucose transporter number and activity (abstract 76). Proc 69th Annu Meet Am Endocr Soc, Indianapolis, June 1987, p. 40.
49 Garvey WT, Hueckst eadt TP, Matthaei S, Olefsky JM: Role of glucose transporters in the cellular insulin resistance of type II non-insulin-dependent diabetes mellitus. J Clin Invest 1988;81: 1528–1536.
50 Karnieli E, Hissin PJ, Simpson IA, et al: A possible mechanism of insulin resistance in the rat adipose cell in streptozotocin-

induced diabetes mellitus: Depletion of intracellular glucose transport systems. J Clin Invest 1981;68:811–814.
51 Karnieli E, Lima FB, Hueckstaedt T: The impact of diabetes on the intrinsic activity of glucose transporters (abstract 53A). 47th Annu Meet ADA, Indianapolis, June 1987. Diabetes 1987; 36(suppl 1):210.
52 Rafaeloff R, Armoni M, Karnieli E: Cycloheximide-modulated glucose transport in insulin-treated diabetic rats (abstract 84A). 47th Annu Meet ADA, Indianapolis, June 1987. Diabetes 1987; 36(suppl 1):333.
53 Scarlett JA, Kolterman OG, Ciaraldi TP, et al: Insulin treatment reverses the post-receptor defect in adipocyte 3-O-methyl glucose transport in type II diabetes mellitus. J Clin Endocrinol Metab 1983;56:1195–1201.
54 Foley JE, Kashiwagi A, Verso MA, et al: Improvement of in vitro insulin action after one month of insulin therapy in obese non-insulin-dependent diabetes. J Clin Invest 1983;72:1901–1909.
55 Garvey WT, Olefsky JM, Griffin RF, et al: The effect of insulin treatment on insulin secretion and insulin action in type II diabetes mellitus. Diabetes 1985;34:222–234.
56 Andrews WJ, Vasques B, Nagulesparan M, et al: Insulin therapy in obese NIDDM induces improvements in insulin action and secretion that are maintained for two weeks after insulin withdrawal. Diabetes 1984;33:634–642.
57 Garvey WT, Hueckstaedt TP, Birnbaum MJ: Pretranslational suppression of an insulin-responsive glucose transporter in rats with diabetes mellitus. Science 1989;245:60–63.

Dr. E. Karnieli
Metabolic Unit
Endocrine Institute
Rambam Medical Center
Haifa 31096 (Israel)

Discussion

H.B. Mortensen, Denmark: Have you studied the glucose transportation using *L*-glucose? You used *D*-glucose and perhaps this transporter cannot recognize the *L*-form.

E. Karnieli, Israel: No, because *L*-glucose has no specific uptake.

Horm Res 1990;33:105–110

0301–0163/90/0334–0105$2.75/0

Role of Growth-Hormone and Insulin-Like Growth-Factor-Binding Proteins

R.L. Hintz

Department of Pediatrics, Stanford University Medical Center, Stanford, Calif., USA

Key Words. Growth hormone · Insulin-like growth factors · Binding proteins · Hormone metabolism

Abstract. Some peptide hormones are associated with specific, high-affinity plasma proteins. The major binding protein (BP) for growth hormone (GH) in humans is a circulating fragment of the GH membrane receptor, consisting of the hydrophilic, extracellular portion of that transmembrane glycoportein. The circulating levels of GH-BP mirror the levels of GH receptors. There are 4 well-characterized insulin-like growth factor (IGF)-BPs. One IGF-binding component in plasma is a fragment of the extracellular portion of the IGF-II/mannose-6-phosphate receptor, analogous to the GH-BP. The 3 other cloned IGF-BPs form a homologous family of proteins with differences in structure, glycosylation and hormonal control that suggest differences in function. The GH- and IGF-BPs play a major role in the metabolism and biological action of these peptide hormones.

Introduction

Many hormones circulate in plasma associated with specific humoral binding proteins (BPs) with high affinity. BPs have been described for thyroid hormones [1], steroid hormones [2], peptide hormones [3, 4] and tissue growth factors [5]. These BPs may play a major role in the metabolism and biological action of these hormones. Potential functions that have been proposed for the BPs include serving as a reservoir of active hormone, increasing the half-life of the hormone and, therefore, maintaining a more constant circulating level, targeting the hormone to specific tissues, modulating the biological action of the hormone, and modifying the metabolism of the hormone (table 1). While data exist to support all of these possible actions of the BPs, absolute proof of the biological *necessity* for the existence of the humoral BPs is still lacking.

Table 1. Potential functions of humoral BPs

1. Reservoir of active hormone
2. Maintenance of a constant level of hormone
3. Targeting to specific tissues
4. Modulation of biological action a) Inhibition b) Potentiation
5. Metabolism of the hormone a) Protection from degradation b) Removal of hormone from tissues

For many years it was commonly thought that the phenomenon of BPs was limited to the thyroidal and steroid hormones. However, in the last decade, solid evidence has developed to support the existence of BPs for both growth hormone- (GH) and the insulin-like growth factors (IGFs), and a great deal of experimental data suggest that these peptide-hormone-BPs may play crucial roles in the biological action of their hormones.

Growth-Hormone-Binding Proteins

The first evidence of a specific plasma BP for GH was presentd by Baumann et al. [4], who showed that ^{125}I-hGH bound specifically to proteins in plasma, and by Herington et al. [6], working with rabbit liver cytosol binding of GH. The main hGH-BP in plasma was shown to have a molecular weight (MW) of about 60 kilodaltons (kDa) and a high affinity ($K_a = 5 \times 10^8$/M). Approximately 40% of hGH circulates complexed to this BP. Early suspicion that this GH-BP and the GH membrane receptor might be related was confirmed when the sequence of the extracytoplasmic portion of the cloned hGH receptor protein was shown to have close homology to the partial amino acid sequence of the high-affinity rabbit GH-BP [7]. This relationship between the high-affinity GH-BP and the GH membrane receptor has now been shown in several species. In man, the evidence suggests that the high-affinity GH-BP is produced by enzymatic cleavage of the hydrophilic extracellular portion of the GH membrane receptor [8], while in other species the GH-BP is produced by alternative splicing of the GH receptor gene [9].

Because of the relationship between the GH receptor and the GH-BP in humans, the ciruculating levels of GH-BP serve as an estimate of the amount of GH receptor. This has been clearly demonstrated in Laron dwarfism, where there are a variety of genetic defects in the GH membrane receptor, all leading to the inability of the GH receptor protein to bind GH. This inability to bind GH is mirrored in the plasma of the patients, where there is a complete absence of the high-affinity GH-BP [10]. The variations of GH-BP in other clinical conditions also seem to indicate the levels of GH membrane receptor, and thus the measurement of GH-BP may be a useful diagnostic tool.

In addition to the high-affinity GH-BP which is clearly related to the GH membrane receptor protein, there is at least one other GH-BP. This GH-BP is larger (MW 100 kDa), has a lower affinity and is present in the plasma of patients with Laron dwarfism [10]. This BP has not yet been completely characterized, but it seems clear that it is not a GH receptor fragment.

The functions of the GH-BPs are unclear. They do increase the half-life and decrease the degradation of GH [11], but also can compete for the binding of GH to the GH membrane receptor [6]. Thus, the net biological effect of the GH-BPs is uncertain.

Insulin-Like Growth-Factor-Binding Proteins

Early observations [12, 13] showed that the 'nonsuppressible insulin-like activity' and 'sulfation factor activity' circulated in plasma as large molecules, but that acid treatment reduced the apparent MW close to that of insulin. This phenomenon was shown to be due to the presence in plasma of specific BPs for the IGFs [3, 14]. On the basis of recombination experiments, it was proposed [15, 16] that the major form of plasma IGF was a 150-kDa complex made up of 3 subunits: IGF-I or -II, a specific IGF-BP and an acid-labile subunit. Smaller amounts of IGF circulated in a 40- to 50-kDa complex made up of IGF-BP and IGF-I or -II, and only a very small proportion of the plasma IGF was free. Work in the past 5 years has verified the correctness of this model of circulating forms of IGF, and has led to the purification, amino acid sequences and cloning of several IGF-BPs. Three structurally distinct but homologous BPs for the IGFs have now been characterized in humans. Because of the confusion of terminology that had built up in the past decade of work on the IGF-BPs, a new nomenclature for these BPs has been proposed [17]. This new terminology, and the myriad of terms it replaces, is summarized in table 2.

IGF-BP1 is the nonglycosylated, GH-independent form of IGF-BP found in high concentrations in amniotic fluid, decidual tissues and in the supernatant of HepG2 cultures. Several groups [18–22] have cloned this BP from different sources, and obtained essentially identical predicted structures for a 25-kDa protein. The IGF-BP1 molecule has no significant homology to IGF receptors, or to any other previously structured protein. Like many receptor proteins, it has cysteine-rich areas that are presumably crucial for the rigid maintenance of the folded three-dimensional structure necessary for ligand binding. Perhaps significantly, IGF-BP1 contains the Arg-Gly-Asp (RGD) sequence that has been implicated in cell-surface interactions in other proteins.

IGF-BP2 was first characterized in the supernatant of rat BRL cells, and has been cloned [23] using probes based on partial amino acid sequence. It has a predicted MW of 31 kDa, and does not seem to be glycosylated. It is homologous to hIGF-BP1, and contains both the cysteine-rich areas and the Arg-Gly-Asp sequence. A human analogue of rIGF-BP2 has recently been cloned and sequenced on the basis of its homology to rat BRL BP [24]. Like IGF-BP1, its synthesis appears to be independent of GH.

IGF-BP3 is the glycosylated, GH-dependent form of IGF-BP found in the 150-kDa IGF complex in plasma. The amino acid sequence predicted from the cloned cDNA structure [25] shows that the MW of the peptide core of IGF-BP3 is 29 kDa, although the apparent MW of the native IGF-BP3 has been estimated as 39–53 kDa. This heterogeneity in MW, and the difference between the predicted and apparent MWs is due to the glycosylation of the IGF-BP3. The predicted amino acid structure contains several potential glycosylation initiation sites, but does not have an Arg-Gly-Asp sequence. The structure of IGF-BP3 has considerable homology to both IGF-BP1 and IGF-BP2, particularly in the cysteine-rich areas, but not to any other known protein. Thus, the IGF-BPs form a closely homologous new family of hormonal ligand-binding proteins.

There are at least 2 other possible members of this IGF-BP family. Hossenlopp et al. [26] have partially sequenced an IGF-BP from cerebrospinal fluid which is homologous, but not identical, to the established IGF-BPs. Ooi and Herington [27] have presented partial sequence data on an IGF-BP of apparent MW of 16 kDa. This amino acid sequence is identical with a region of IGF-BP3, but does not seem to arise via proteolytic cleavage of IGF-BP3. It is possible that this form of IGF-BP is an alternative splicing of the IGF-BP3 gene. Until definitive information about other forms of IGF-BPs is available, their place in the family of IGF-BPs cannot be assigned with confidence, and proliferation of terminology should be avoided.

There is at least 1 form of plasma IGF-BP that is not homologous to the emerging IGF-BP family. Keiss et al. [28] have demonstrated that a high-MW component of IGF-specific binding is due to a circulating fragment of the IGF-II/mannose-6-phosphate receptor. This appears to be analogous to the high-affinity GH binding due to the extracellular portion of the GH membrane receptor.

The major normal form of circulating IGF is the GH-dependent 150-kDa complex composed of 3 subunits: IGF-I or IGF-II peptide, glycosylated IGF-BP3 and an acid-labile subunit of 90 kDa. This acid-labile subunit has been characterized by Baxter [29], but no sequence information is yet available. Thus, it is uncertain whether this subunit has any homology to the IGF-BPs or any other protein. What is clear is that this acid-labile component will recognize and bind only to IGF already complexed to IGF-BP3, and not IGF or IGF-BP3 alone. Very little IGF-BP3 is found outside of the 150-kDa IGF complex, and conversely no IGF-BP1 or IGF-BP2 is found in the 150-kDa complex [30–31].

Table 2. Nomenclature of the IGF-BPs

	Previous terms
IGF-BP1	Amniotic fluid BP Placental protein 12 α_1-PEG GH-independent BP BP28, BP26, BP25 IBP-1
IGF-BP2	IBP-2 BRL 3a IGF BP MDBK BP
IGF-BP3	GH-dependent BP Acid-stable subunit of the 140 K IGF complex BP53 BP29

Role of Insulin-Like Growth-Factor-Binding Proteins

The IGF peptides bound to BPs have a vastly increased half-life in plasma [14]. This is particularly true of the 150-kDa complex composed of IGF peptide, IGF-BP3 and the acid-labile component, which represents the vast majority of plasma IGF. Most estimates of the half-life of the 150-kDa IGF complex are in terms of hours rather than the few minutes of free IGF peptides. One consequence of this long half-life is that the total circulating level of IGF peptides in plasma (IGF-I + IGF-II) is close to 1,000 ng/ml, which is 100- to 1,000-fold higher than most peptide hormones. Another consequence is that there is little variation in plasma IGF levels throughout the day. Mathematical modeling [32] demonstrates that in most circumstances the circulating level of IGF is more dependent on the level of IGF-BPs than on the synthesis rate of the IGF peptides.

The IGF-BP levels are under the control of specific factors. The level of IGF-BP3, and the 150-kDa IGF complex, appears to be mainly under the control of GH, with low levels of IGF-BP3 in GH deficiency and high levels of IGF-BP3 in acromegaly [30]. The levels of IGF-

Table 3. Characteristics of the IGF-BPs

BP	Size	Glycosylation	Regulation
IGF-BP1	25 kDa	no	GH-independent insulin-dependent fetal
IGF-BP2	31 kDa	no	GH-independent insulin-dependent fetal
IGF-BP3	29 kDa	yes	GH-dependent insulin-independent adult life

BP1 are under control of insulin, with low levels of IGF-BP1 in the presence of high levels of insulin, and vice versa [33]. Thus, in situations of anabolism which are typically associated with high GH and insulin levels, formation of BP3 and the 150-kDa IGF complex is favored. On the other hand, in situations of catabolism associated with low insulin levels and either low GH levels or a block in GH action, the formation of BP1 and BP2 and the 40-kDa IGF complex are promoted. Teleologically, this would imply that the 150-kDa IGF complex and IGF-BP3 are involved with anabolism and growth, but as yet there is no proof of this.

Just as IGF peptides have been shown to be produced in a variety of tissues, suggesting an autocrine or paracrine role of IGF peptides, IGF-BPs have also been shown to be made by many tissues including decidual cells [12], liver cells [18], fibroblasts [34] and breast cancer cells [35]. It is too early to generalize with confidence, but it appears that the phenomenon of cells making IGF-BPs is at least as common as cells making IGF peptides, and in many cases where it has been studied carefully the cells are making larger molar amounts of IGF-BPs than of IGF peptides. The types of IGF-BP made vary from cell to cell and are under humoral control [34].

There is evidence that the IGF-BPs can have both inhibitory and stimulatory effects on the action of IGF peptides. Bagley et al. [36] have shown that inhibition of the action of IGF peptide analogues is directly related to their ability to bind to IGF-BPs. The more IGF peptide that is bound the less biological action is seen. These data, and others, are consistent with the hypothesis that only the free IGF peptide is biologically active. On the other hand, other workers [37] have shown a potentiation of IGF action in the presence of IGF-BPs. The solution of this apparent paradox may lie in differing molecular forms of IGF-BP [38], with some forms of IGF-BPs being inhibitory and some stimulatory. Alternatively, the IGF-BPs may be synergistic with IGF peptide action at low molar ratios but inhibitory when present in excess [39]. We have essentially no data as yet on the effects of the 150-kDa complex, which is the major normal circulating form of IGF, and seems to be correlated with the anabolic status of the whole organism.

Conclusions

Just as is true for thyroidal hormones and steroid hormones, the peptide hormones GH and the IGF peptides have specific BPs. The major BP for GH in humans is a ciruculating fragment of the GH membrane receptor, consisting of the hydrophilic, extracellular portion of that transmembrane glycoprotein. The formation of this GH receptor fragment is not well characterized in man, and at least one other GH-BP remains to be fully defined.

There are now 4 IGF-BPs that have been well characterized. One minor IGF-binding component in plasma is a fragment of the extracellular portion of the IGF-II/mannose-6-phosphate receptor, analogous to the GH-BP. The 3 other cloned IGF-BP proteins form a homologous family of proteins with differences in structure, glycosylation and hormonal control that suggest differences in function (table 3). There may be other members of this IGF-BP family. The IGF-BPs clearly lengthen the half-life of the IGF peptides, and change the metabolism of IGF. The BPs also affect biological action by the modulation of hormonal binding to cell surface receptors, and both inhibitory and stimulatory effects of the IGF-BPs on IGF action have been described. At this early stage of research, we only have glimpses of a whole new level of control of IGF action. Further studies with pure materials and well-defined conditions may show that the BPs play a pivotal role in the control of anabolism and growth.

References

1 Robbins J, Cheng S-Y, Gershengorn MC, et al: Thyroxine transport proteins of plasma: Molecular properties and biosynthesis. Recent Prog Horm Res 1978;34:477–519.
2 Siiteri P, Marai JT, Hammond GC, et al: The serum transport of steroid hormones. Recent Prog Horm Res 1982;38:457–510.

3 Hintz RL, Liu F: Demonstration of specific plasma protein binding sites for somatomedin. J Clin Endocrinol Metab 1977; 45:988–995.
4 Baumann G, Stolar MW, Amburn K, et al: A specific growth-hormone-binding protein in human plasma: Initial characterization. J Clin Endocrinol Metab 1986;62:132–141.
5 James R, Bradshaw RA: Polypeptide growth factors. Annu Rev Biochem 1984;53:259–292.
6 Herington AC, Ymer S, Stevenson J: Identification and characterization of a specific binding protein for growth hormone in normal human sera. J Clin Invest 1986;77:1817–1823.
7 Leung DW, Spencer SA, Cachianes G, et al: Growth hormone receptor and serum binding protein: Purification, cloning and expression. Nature 1987;330:537–543.
8 Trivedi B, Daughaday WH: Release of growth hormone binding protein from IM-9 lymphocytes by endopeptidase is dependent on sulfhydryl group inactivation. Endocrinology 1988;123: 2201–2206.
9 Smith WC, Kuniyoshi J, Talamantes F: Mouse serum GH binding protein has GH receptor extracellular and substituted transmembrane domains. Mol Endocrinol 1989;3:984–990.
10 Baumann G, Winter RJ, Shaw M: Absence of the plasma growth hormone-binding protein in Laron-type dwarfism. J Clin Endocrinol Metab 1987;65:814–816.
11 Baumann G, Amburn KD, Buchanan TA: The effect of circulating growth hormone binding protein on metabolism, clearance, distribution and degradation of human growth hormone. J Clin Endocrinol Metab 1987;64:657–660.
12 Jacob A, Hauri CH, Froesch ER: Non-suppressible insulin-like activity in human serum. 3. Differentiation of two distinct molecules with non-suppressible ILA. J Clin Invest 1968;47:2678–2685.
13 van Wyk JJ, Underwood LE, Hintz RL, et al: The somatomedins: A family of insulin-like hormones under growth hormone control. Recent Prog Horm Res 1974;30:259–318.
14 Kaufman U, Zapf J, Torretti B, et al: Demonstration of a specific serum carrier protein of non-suppressible insulin-like activity in vivo. J Clin Endocrinol Metab 1977;44:160–166.
15 Hintz RL, Liu F: Human somatomedin plasma binding proteins; in Giordano G, Van Wyk JJ, Minuto F (eds): Somatomedins and Growth. New York, Academic Press, 1979, pp 143–148.
16 Furlanetto RW: The somatomedin C binding protein: Evidence for a heterologous subunit structure. J Clin Endocrinol Metab 1980;51:12–19.
17 Ballard FJ, Baxter RC, Binoux M, et al: Report on the nomenclature of the IGF binding proteins. J Clin Endocrinol Metab 1990;70:817–818.
18 Lee Y-L, Hintz RL, James PM, et al: Insulin-like growth factor binding protein cDNA from human HEP G2 hepatoma cells: Predicted protein sequence suggests an IGF binding domain different from those of IGF-I and IGF-II receptors. Mol Endocrinol 1988;2:404–411.
19 Cubbage ML, Suwanichkul A, Powell DR: Structure of the human chromosomal gene for the 25 kilodalton insulin-like growth factor binding protein. Mol Endocrinol 1989;3:846–851.
20 Brinkman A, Groffen C, Kortleve DJ, et al: Isolation and characterization of a cDNA encoding the low molecular weight insulin-like growth factor binding protein. EMBO J 1988;7:2417–2423.
21 Julkunen M, Koistinen R, Aalto-Setala K, et al: Primary structure of human IGF binding protein/placental protein 12, and tissue-specific expression of its mRNA. FEBS Lett 1988;236: 295–302.
22 Brewer MT, Stetler GL, Spuires CH, et al: Cloning, characterization, and expression of a human insulin-like growth factor binding protein. Biochem Biophys Res Commun 1988;152:1289–1297.
23 Rechler MM, Yang YW-H, Brown AL, et al: Molecular characterization of insulin-like growth factor binding proteins in the rat; in Drop SLS, Hintz RL (eds): The Insulin-Like Growth Factor Binding Proteins. Amsterdam, Excerpta Medica, 1989, pp 133–142.
24 Binkert C, Landwehr J, Mary JL, et al: Cloning, sequence analysis and expression of a cDNA encoding a novel insulin-like growth factor binding protein (IGFBP-2). EMBO J 1989;8: 2497–2502.
25 Wood WI, Cachianes G, Henzel WJ, et al: Cloning and expression of the growth hormone-dependent insulin-like growth factor-binding protein. Mol Endocrinol 1989;2:1176–1185.
26 Hossenlopp P, Seurin D, Segovia-Quinson B, et al: Identification of an insulin-like growth factor binding protein in human cerebrospinal fluid with a selective affinity for insulin like growth factor II: FEBS Lett 1986;208:439–444.
27 Ooi GT, Herington AC: Recognition of insulin-like growth factor serum binding proteins by an antibody raised against a specific insulin-like growth factor inhibitor. Biochem Biophys Res Commun 1988;156:783–791.
28 Keiss W, Greenstein LA, White RM, et al: Type II insulin-like growth factor receptor is present in rat serum. Proc Natl Acad Sci USA 1987;84:7720–7724.
29 Baxter RC: Characterization of the acid-labile subunit of the growth hormone-dependent insulin-like growth factor binding protein complex. J Clin Endocrinol Metab 1988;67:265–272.
30 Hardouin S, Hossenlopp P, Segovia B, et al: Heterogeneity of insulin-like growth factor binding proteins and relationships between structure and affinity. I. Circulating forms in man. Eur J Biochem 1987;170:121–132.
31 Liu F, Powell DR, Hintz RL: Characterization of insulin-like growth factor binding proteins in human serum from patients with chronic renal failure. J Clin Endocrinol Metab 1990;70: 620–628.
32 Blum WF, Ranke MB, Kietzmann K, et al: Excess of IGF-binding proteins in chronic renal failure: Evidence for relative GH resistance and inhibition of somatomedin activity; in Drop SLS and Hintz RL (eds): The Insulin-Like Growth Factor Binding Proteins. Amsterdam, Excerpta Medica, 1989, pp 93–102.
33 Suikkari AM, Koivisto VA, Rutanen EM, et al: Insulin regulates the serum levels of low molecular weight insulin-like growth factor-binding protein. J Clin Endocrinol Metab 1988;66:266–272.
34 Conover CA, Liu F, Powell D, et al: Insulin-like growth factor binding proteins from cultured human fibroblasts: Characterization and hormonal regulation. J Clin Invest 1989;83:852–859.
35 De Leon DD, Wilson DM, Bakker B, et al: Characterzation of insulin-like growth factor binding proteins from human breast cancer cells. Mol Endocrinol 1989;3:567–574.

36 Bagley CJ, May BL, Szabo L, et al: A key functional role for the insulin-like growth factor 1 N-terminal pentapeptide. Biochem J 1989;259:665–671.
37 Blum WF, Jenne EW, Reppin F, et al: Insulin-like growth factor I binding protein complex is a beter mitogen than free IGF-I. Endocrinology 1989;125:766–772.
38 Busby WH, Hossenlopp P, Binoux M, et al: Purified preparations of the amniotic fluid-derived insulin-like growth factor-binding protein contain multimeric forms that are biologically active. Endocrinology 1989;125:773–777.
39 DeMellow JSM, Baxter RC: Growth hormone-dependent insulin-like growth factor binding protein both inhibits and potentiates IGF-I stimulated DNA synthesis in human skin fibroblasts. Biochem Biophys Res Commun 1988;156:199–204.

Dr. R.L. Hintz
Department of Pediatrics
Stanford University Medical Center
Stanford, CA 94305 (USA)

Discussion

Z. Hochberg, Israel: What are the implications of the BP on the measurements of serum IGF-I?

R. Hintz, USA: For a while I thought that God had invented BPs to confuse IGF-I determinations. The presence in plasma of free BPs makes it difficult to measure IGF in whole serum and one has to get rid of the free BPs. In addition, if one wishes to measure the total amount of IGF that is present in plasma, you must break the complex apart by acid or some other means.

Z. Laron, Israel: We have recently reported that a large part of family members of Laron-type dwarfs have subnormal serum GH-BP values, and we consider this to be a marker for the diagnosis of heterozygocity for this disease [Laron et al., Acta Endocrinol 1989; 121:603–608].

Horm Res 1990;33:111–115

0301–0163/90/0334–0111$2.75/0

In vitro Insulin-Like Growth Factor I Interaction with Cartilage Cells Derived from Postnatal Animals

D. Demarquay, M.F. Dumontier, L. Tsagris, J. Bourguignon, V. Nataf, M.T. Corvol

Inserm U 30, Unité de Recherches en Biologie et Pathologie de la Croisssance et du Développement, Hôpital des Enfants-Malades, Paris, France

Key Words. IGF-I · Chondrocyte culture · IGF-I, local production · Collagen · RNA

Abstract. This paper reports data on the in vitro effects of insulin-like growth factor I (IGF-I) and basic fibroblast growth factor (bFGF) on the phenotypic expression of epiphyseal chondrocytes grown in serum-free (SF) culture medium. bFGF mostly stimulates chondrocyte DNA and inhibits sulfated proteoglycan synthesis and type II collagen mRNA. On the contrary, IGF-I is poorly mitogenic but strongly stimulates protein synthesis and type II collagen mRNA. In addition, IGF-I prevents the expression of type I collagen gene. Lastly, chondrocytes cultured in SF medium are able to locally produce IGF-I peptides. In conclusion, IGF-I and bFGF have opposite effects on the phenotypic expression of chondrocytes in vitro: bFGF is mostly mitogenic and IGF-I appears to be a differentiating factor.

Introduction

The purpose of this study was to try to better understand the mode of action of insulin-like growth factor I (IGF-I) on chondrocyte phenotype during postnatal cell division. Chondrocyte phenotype has been characterized in terms of proteoglycan and collagen synthesis. The latter parameter is classically used to follow the process of chondrogenesis in embryos [1, 2]. Chondrogenesis involves the differentiation of prechondrogenic mesenchymal cells into mature chondroblasts. During this process a transition from type I to type II collagen synthesis occurs. There is a total cessation of type I collagen peptide synthesis but the differentiated chondroblasts still contain the message.

This process was demonstrated by Kosher et al. [2] using a limb-bub culture system which allows maturation of prechondroblasts in vitro as followed by the progressive elevation of type II collagen mRNA. But at the same time, these cells which stopped to synthesize type I collagen peptide still produce type I mRNA. These data suggest that during chondroblast differentiation there must be some translational control involved in the prevention of type I collagen synthesis besides the strong transcriptional control of type II collagen gene expression. The nature of the controlling factors or hormones and their mode of action are not known.

During postnatal skeletal growth or repair the modulation of chondrocyte phenotype is still under investigation. First of all the cells present in epiphyseal cartilage of long bones (articular and growth plate) are well-differentiated cartilage cells and mainly synthesize type II collagen. During postnatal prepubertal and pubertal life, epiphyseal chondrocytes start to proliferate with a limited number of cell divisions before hypertrophy, death and calcification. During this process as well as during cartilage repair following bone fracture, chondrocytes still synthesize type II collagen and start to express new collagen genes such as type XI [3] and type X collagen [4]. When cartilage calcifies, hypertrophic chondrocytes were also shown to accumulate intra- and extracellularly the C propeptide of type II collagen, previously called chondrocalcin [5, 6].

Our question is: among all the growth factors which may be involved in this maturation process, in what extent and at what level IGF-I by itself is able to modulate chondrocyte phenotype?

Material and Methods

Chondrocyte Culture

In vitro, differentiated chondrocytes isolated by collagenase digestion of epiphyseal cartilage from postnatal mammals are able to enter into a proliferative phase in certain culture conditions, and particularly in the presence of growth-promoting factors. In that case chondrocyte phenotype is changing [7, 8]. The cells gain a less-differentiated stage as evaluated by their ability to synthesize again type I collagen peptide.

We have used such a culture system in which chondrocytes from the resting zone of growth plate cartilage or epiphyseal chondrocytes from articular cartilage were grown in serum-free (SF) medium in the presence of basic fibroblast growth factor (bFGF; 1–50 ng/ml) alone or combined with IGF-I (1–40 ng/ml). (bFGF has been shown to be a growth-promoting factor for mesenchymal-derived cells [9].) The chondrocyte phenotype was followed by morphological and metabolic parameters.

^{35}S-Sulfate Incorporation into Chondrocyte Proteoglycans

Newly synthesized radiolabeled proteoglycans secreted by cultured chondrocytes were evaluated after ^{35}S-sulfate incorporation as previously described [10].

Chondrocyte RNA Extraction: Analysis of Collagen mRNA and Collagen Peptides

Total RNA of cultured cells was extracted and analyzed by dot blot or Northern blot hybridization as described [11] using cDNA probes specific for type II α_1II and type I α_2I collagen chains generously provided by Dr. Ramirez (USA).

Isolation of newly synthesized radiolabeled collagenous proteins was performed after ^{3}H-proline incorporation followed by gel electrophoresis of culture medium and cell lysate as described [11].

Acid Extraction and Chromatography of Immunoreactive IGF-I Fractions Isolated from Chondrocyte Culture Medium

Culture medium was acid-extracted with 1 *M* acetic acid prior to chromatography on Biogel P100. The 6- to 25-kilodalton (kDa) fraction was recovered and rechromatographed on Biogel P30. IGF-I activity was evaluated by radioimmunoassay measured in each eluate. Further characterization of IGF-I fractions was performed by isoelectrofocalization and high-performance liquid chromatography.

Newly synthesized immunoreactive IGF-I fractions by chondrocytes were demonstrated after pulse chase experiments with ^{35}S-cysteine followed by IGF-I immunoprecipitation of acid-extracted cell lysates and corresponding culture medium, chromatographed as described above.

Results and Discussion

Morphology

In the present experimental conditions, bFGF induces DNA synthesis and cell multiplication of cultured epiphyseal chondrocytes but it also modulates chondrocyte metabolism in a dedifferentiating pathway.

When incubated for 3 days in the presence of bFGF alone, the rounded shape of the resting chondrocytes or the articular cells grown at high-density culture is rapidly changed. Chondrocytes become elongated and fibroblast-like. The addition of IGF-I does not modify the morphology unless bFGF is removed from the culture medium, thus demonstrating the reversible effect of bFGF on chondrocyte morphology.

^{35}S-Sulfate Incorporation into Chondrocyte Proteoglycans

In the presence of bFGF alone, radiolabeled proteoglycans newly synthesized by chondrocytes are significantly decreased compared with controls. Such an inhibitory effect of bFGF was suggested to be the consequence of the effect of growth factors on the induction of the cells to enter the S phase and to synthesize DNA, thus blocking the expression of cellular protein genes. When added together with bFGF, IGF-I stimulates ^{35}S-sulfated proteoglycan synthesis, an effect which is augmented when increasing the concentration of bFGF. A high concentration insulin (1–5 µg/ml) produces similar effects as IGF-I (10–40 ng/ml), thus confirming the probable mediation of such an effect through the IGF-I receptors.

Cartilage has been suggested to be a target tissue for bFGF which strongly stimulates DNA synthesis in chondrocytes in vitro [12]. In addition, structural studies and recent cloning of the bFGF gene [13] have shown its homology with many other growth factors. It is now likely that bFGF may be produced by cartilage cells or by endothelial cells from the microvessels heading chondrocytes in the resting zone of growth plate cartilage or even by bone cells surrounding the mineralized matrix of the epiphyseal plate [14]. bFGF could thus act either as an autocrine or paracrine growth factor in growth plate cartilage. But the mechanism of action of this factor remains unclear. The widespread distribution of bFGF as well as the wide range of cell types on which it acts suggest that its growth-promoting effect is not tissue-specific [for review, see ref. 15].

In fact, bFGF has been proposed to act as a competence factor, as other factors such as platelet-derived growth factor and epidermal growth factor [16], while another group of factors is responsible for the progression of normal diploid cells committed to divide by brief exposure to competence factor [17, 18]. IGF-I was proposed to be one major progression factor for cartilage growth with specific stimulating effect on cartilage sulfation.

This is well demonstrated in the present study, in which bFGF stimulates chondrocyte DNA and IGF-I thus appears to be able either to prevent or to reverse the inhibitory effect of bFGF on proteoglycan synthesis.

Such an increased sensitivity to IGF-I by proliferative chondrocytes as compared with resting chondrocytes has been previously shown by Trippel et al. [19] using a similar system of chondrocyte culture.

Collagen Gene Expression by Chondrocytes Exposed to bFGF Alone or with IGF-I

In the presence of bFGF alone we observe a significant decrease in type II mRNA level but no modification of type I mRNA. When IGF-I (10–40 ng/ml) is added to the culture medium, type II mRNA level is stimulated but again no significant variation of type I mRNA is observed. The effect of IGF-I is reproduced by a high concentration of insulin [20].

In addition, both bFGF and IGF-I, again in an opposite direction, were shown to modulate the production of collagen peptides in chondrocytes. The amount of type I collagen chains is strongly increased in the presence of bFGF alone. The addition of IGF-I almost completely suppresses the production of type I collagen peptide.

Thus, collagen type I synthesis appears to be mostly regulated at the posttranscriptional level, as was previously suggested by Willis and Liberti [21] and by Bennett and Adams [22].

Because relaxin has been shown to contain structural homologies with insulin and IGFs and is one of the insulin family peptides [23], its effect on chondrocyte phenotype was compared with the effect of IGF-I [11]. This peptide is produced by ovaries during pregnancy and is known to induce relaxation of uterine muscle at parturition. When added to chondrocytes grown in SF medium, with 10 ng/ml bFGF, type I collagen mRNA level increases and type II mRNA level is not changed. At the peptide level, type I collagen is also increased since type II collagen content is not modified. This effect is reversible when relaxin is removed from the culture medium.

Thus, relaxin appears to act on chondrocyte phenotype in a completely different way than IGF-I or insulin demonstrating that growth factors from the same family of peptides could have completely different target effects. Interestingly, many years ago it has been suggested that the anatomical modifications of pubic cartilage observed before and at parturition could be related to relaxin's effect. Cartilage is transformed in what is called 'fibrocartilage' with alteration of collagen content. Until now the possible role of relaxin on this transformation has not been demonstrated. The present data could, at least partly, explain how relaxin may alter cartilage cell metabolism and differentiation.

Production of IGF-I Peptides by Cultured Chondrocytes

The last part of the present data concerns the ability of cultured chondrocytes exposed to bFGF alone to produce IGF-I peptides in the absence of any hormone added.

Fractions with IGF-I and IGF-II activities have been recently shown to be released in culture medium of rabbit articular chondrocytes [24].

In the present experimental conditions an immunoreactive IGF-I fraction was also shown to be released into chondrocyte culture medium. When rechromatographed on Biogel P30, this fraction appears to be heterogeneous in terms of molecular weight, mostly composed of 3 fractions of 15–18, 10–12 kDa and 6–8 kDa, respectively. After pulse chase experiments of ^{35}S-methionine and ^{35}S-cysteine followed by immunoprecipitation. The high-MW fraction was shown to be partly converted into the 6- to 8-kDa fraction, thus resulting in a predominant accumulation of the 10- to 12 kDa fraction in culture medium at equilibrium [D. Demarquay et al., unpubl. data].

In addition, IGF-I activity released in culture medium is increased when increasing the bFGF concentration used, still containing a major proportion of 10- to 12-kDa peptides.

Conclusions

The present data show that: (1) IGF-I and bFGF have opposite or likely complementary effects concerning DNA synthesis and specific protein synthesis by cultured epiphyseal chondrocytes; (2) bFGF, which mostly acts as a mitogenic factor, is also modulating chondrocyte phenotype in a dedifferentiating pathway; (3) exogenous IGF-I when added together with exogenous bFGF can prevent the dedifferentiating effect of bFGF by acting at the translational level and probably also at the transcriptional level of chondrocyte collagen gene expression, and (4) it is of note that the chondrocyte itself can react against the induced dedifferentiating effect of bFGF by producing its own 'IGF-I-differentiating' factors.

Whether or not this newly synthesized IGF-I peptide mostly represented by the 10- to 12-kDa IGF-I compounds may act as an autocrine factor for these cells and play an important role in the final differentiating process of chondrocytes remains to be studied. More precisely, does IGF-I modulate minor collagen gene expression such as type X, type XI and/or type II C propeptide? This effect of IGF-I as a stimulating factor of late differentiation of mesenchymal-derived cells has been first proposed by Zezulak and Green [25] and has been well demonstrated in adipocytes [25, 26] and muscle cells [27]. Further studies are necessary in order to characterize its effect on cartilage cell phenotype.

References

1 Von der Mark K, von der Mark H: Immunological and biochemical studies of collagen type transition during in vitro chondrogenesis of chick limb mesodermal cells. J Cell Biol 1977;73:736–747.

2 Kosher RA, Kulyk WM, Gay SW: Collagen gene expression during limb cartilage differentiation. J Cell Biol 1986;102:1151–1156.

3 Vasios G, Nishimura I, Konomi H, et al: Cartilage type IX collagen-proteoglycan contains a large amino-terminal globular domain enroded by multiple exons. J Biol Chem 1988;263:2324–2329.

4 Grant WT, Wang GJ, Balian G: Type X collagen synthesis during endochondral ossification in fracture repair. J Biol Chem 1987;262:9844–9849.

5 Van der Rest M, Rosenberg LC, Olsen BR, et al: Chondrocalcin is identical to C propeptide of type II procollagen. Biochem J 1986;297:923–925.

6 Hinek A, Reiner A, Poole AR: The calcification of cartilage matrix in chondrocyte culture: Studies on the C-propeptide of type II collagen (chondrocalcin). J Cell Biol 1987;104:1435–1441.

7 Benya PD, Brown PD: Modulation of the chondrocyte phenotype in vitro; in Kuettner KE, Schleyerbach R, Hascall VC (eds): Articular Cartilage Biochemistry. New York, Raven Press, 1986, pp 219–223.

8 Mayne R, Vail MS, Mayne PM, Miller EJ: Changes in the type of collagen synthesized as clones of chick chondrocytes grow and eventually lose division capacity. Proc Natl Acad Sci USA 1976; 73:1674–1681.

9 Prins AP, Lipman JM, Sokoloff L: Effect of purified growth factors on rabbit articular chondrocytes in monolayer culture. II. Arthritis Rheum 1982;25:1228–1238.

10 Corvol MT, Carrascosa A, Tsagris L, et al: Evidence for a direct in vitro action of sex steroid on rabbit cartilage cells during skeletal growth. Endrocrinology 1987;120:1422–1429.

11 Bonaventure J, de la Tour B, Tsagris L, et al: Effect of relaxin on the phenotype of collagens synthesized by cultured rabbit chondrocytes. Biochim Biophys Acta 1988;972:209–220.

12 Sachs BL, Goldberg M, Moskowitz RW, et al: Response of articular chondrocytes to pituitary fibroblast growth factor (FGF). J Cell Physiol 1982;112:51–59.

13 Esch F, Baird A, Ling N, et al: Primary structure of bovine pituitary basic fibroblast growth factor (bFGF) and comparison with the aminoterminal sequence of bovine brain acidic FGF. Proc Natl Acad Sci USA 1985;82:6507–6511.

14 Hauschka PV, Mavrakas AE, Iafrati MD, et al: Growth factors in bone matrix: Isolation of multiple types by affinity chromatography on heparin-Sepharose. J Biol Chem 1986;261:12665–12674.

15 Gospodarowicz D, Ferrara N, Schweigerer L, et al: Structural characterization and biological functions of fibroblast growth factor. Endocr Rev 1987;8:95–113.

16 Jiminez de Asua L, O'Farell MK, Clinigan D, et al: Temporal sequence of hormonal interactions during the prereplicative phase of quiescent 3T3 fibroblasts. Proc Natl Acad Sci USA 1977;74:3845–3849.

17 Stiles CD, Capone GT, Scher CD, et al: Dual control of cell growth by somatomedins and platelet-derived growth factor. Proc Natl Acad Sci USA 1979;76:1279–1283.

18 Adashi EY, Resnick CE, Svoboda ME, et al: A novel role for somatomedin C in the cytodifferentiation of the ovarian granulosa cell. Endocrinology 1984;115:1227–1229.

19 Trippel SB, Corvol MT, Dumontier MF, et al: Effect of somatomedin C/insulin like growth factor I and growth hormone on cultured growth plate and articular chondrocytes. Pediatr Res 1989;25:76–82.

20 Nataf V, Tsagris L, Dumontier MF, et al: Modulation of sulfated proteoglycan synthesis and collagen gene expression by chondrocytes grown in the presence of bFGF alone or combined with IGF_1. Reprod Nutr Dev 1990;30:17–21.

21 Willis DH Jr, Liberti JP: Post-receptor actions of somatomedin on chondrocyte collagen biosynthesis. Biochim Biophys Acta 1985;844:72–80.

22 Bennett VD, Adams SL: Characterization of the translational control mechanism preventing synthesis of α_2 1 collagen in chicken vertebral chondroblasts. J Biol Chem 1987;262:14806–14814.

23 Bell GI, Merryweather JP, Sanchez-Pescador R, et al: Sequence of a cDNA clone encoding human preproinsulin-like growth factor II. Nature 1984;310:777–781.

24 Froger-Gaillard B, Hossenlop P, Adolphe M, et al: Production of insulin like growth factors and their binding proteins by rabbit articular chondrocytes: Relationship with cell multiplication. Endocrinology 1989;124:2365–2372.

25 Zezulak KM, Green H: The generation of insulin-like growth factor 1-sensitive cells by growth hormone action. Science 1986; 233:551–553.

26 Doglio A, Dani C, Fredrikson G, et al: Acute regulation of insulin-like growth factor 1 gene expression by growth hormone during adipose cell differentiation. EMBO J 1987;6:4011–4016.

27 Florini JR, Magri KA: Effects of growth factors on myogenic differentiation. Am J Physiol 1989;256:C701–C711.

Dr. M.T. Corvol
INSERM U 30
Unité de Recherches en Biologie
et Pathologie de la Croissance et du Développement
Hôpital des Enfants-Malades
Tour Technique 6e étage
149, rue de Sèvres
F–75743 Paris, Cedex 15 (France)

Discussion

Z. Laron, Israel: In your chondrocyte cultures did you observe that IGF-I acts alone without the need of growth hormone or insulin?

M.T. Corvol, France: We only saw the effect of IGF-I in the absence of growth hormone and insulin. IGF-I by itself acts on the chondrocytes at a different level in the absence of growth hormone or any other hormones and in the absence of insulin. Apparently insulin only mimics the effect of IGF-I in these cells which have been shown to contain IGF-I receptor and not insulin receptors.

L. van den Brande, The Netherlands: Have you had the chance to look at the time relationship of adding FGF and IGF at different times in your incubation system to see whether you perhaps get different effects?

M.T. Corvol: In order to characterize the effect of IGF-I on chondrocyte differentiation we first added FGF and then IGF-I. We tried to use SF medium and to mix FGF with IGF-I in order to get good mRNAs and not the differentiation. But until now the peptide concentrations used seem to be too high. Apparently we need other factors as well.

G. Werther, Australia: Is it possible that the effect of FGF is on a subpopulation of slightly less differentiated cells? Can you be sure that the two hormones were acting on exactly the same cells?

M.T. Corvol: I am not sure that the cells are exactly the same as we have a mixed population. We are sure that we have only chondrocytes but the stage of the cell cycle is not the same in all. Therefore, your suggestion is possibly correct.

Horm Res 1990;33:116–120

0301–0163/90/0334–0116$2.75/0

Status Report on the US Human Growth Hormone Recipient Follow-Up Study

J.L. Mills[a], *J. Fradkin*[a], *L. Schonberger*[b], *W. Gunn*[b], *R.A. Thomson*[c], *J. Piper*[d], *D. Wysowski*[e], *P. Brown*[a]

[a] National Institutes of Health, USPHS, Bethesda, Md.; [b] Centers for Disease Control, Atlanta, Ga.; [c] Westat, Inc., Rockville, Md.; [d] Vanderbilt University, Nashville, Tenn.; [e] Food and Drug Administration, USPHS, Rockville, Md., USA

Key Words. Human growth hormone · Creutzfeldt-Jakob disease · Pituitary growth hormone · Leukemia · (US) cohort study

Abstract. Three reported cases of Creutzfeldt-Jakob disease (CJD) in young adults who had received human growth hormone (hGH) raised concerns that pituitary-derived GH had been contaminated. Subsequently reported cases have confirmed this suspicion. The US Public Health Service is conducting an investigation to determine the extent of the problem of CJD in recipients of National Hormone Pituitary Program (NHPP) GH. In addition, other possible adverse effects of GH use including leukemia are being investigated. The design, conduct and current status of the study are the subject of this report. Interview data are now available on 5,240 of the 6,284 subjects treated by the NHPP for growth problems. Analysis is underway.

Introduction

Creutzfeldt-Jakob disease (CJD) is a uniformly fatal neurological disease characterized by progressive dementia associated with cerebellar dysfunction, myoclonus and characteristic autopsy findings of diffuse neuronal loss, astrocytosis and spongiosis [1]. It is a disease of older adults. Of over 3,000 cases reported in the medical literature or referred to the Laboratory of Central Nervous System Diseases (NINDS) for study, only 9 occurred before age 30 [2].

Thus, it was a matter of great concern in 1985 when 3 cases of CJD were found to have occurred in young adults who had received pituitary-derived human growth hormone (hGH) through the National Hormone and Pituitary Program (NHPP) sometime between 1963 and 1980 [3–6]. This discovery raised concerns over the safety of pituitary-derived hGH. Distribution of NHPP hormone was halted by NIDDK and the FDA recommended that the sale and use of commercial hGH be discontinued in April 1985. Representatives of the NIH, CDC, FDA, the Pediatric Endocrinology community and interested outside organizations met in May of the same year; they recommended that an epidemiologic study be conducted to assess the risk of CJD and other complications of hGH therapy in subjects treated through the NHPP program.

This paper will describe the design, conduct and current status of that study.

Identifying GH Recipients

The first goal of the US hGH Recipient Follow-Up Study was to determine the health status of all patients who received NHPP hGH for the treatment of growth failure over the 22-year period that hGH was distributed (1963–1985). Patients who received hGH exclusively from pharmaceutical companies were not included because the study group had no authority to obtain treatment records or other data on them.

The first challenge the study faced was to determine how many patients actually received hGH and who they were. Initially, the NHPP had kept records of patients' birthday or age, sex and name. In 1973, initials were substituted for names, and in 1977 a code known only to the treating physician was substituted for initials. Because requests for renewal of hGH were filed annually, many patients appeared numerous times, sometimes under their actual names, sometimes under initials and sometimes under codes. In all, there were over 26,000 applications and it appeared that over 10,000 patients had received hGH.

It was necessary to contact over 200 treatment centers to determine the actual number of hGH recipients, why they received hGH, and how they could be contacted. Fortunately, the treating physicians provided invaluable assistance by completing abstracts on each of their patients.

The information provided by the treating physicians made it possible to edit the 26,000 plus applications for hGH. After duplicate applications were excluded, there were 8,139 potential subjects identified from the NHPP records (table 1). Eighteen additional recipients of NHPP hGH were identified from other sources during the study. The study investigators found that 762 NHPP patient listings could not be identified from the treatment center listings. There were 70 potential subjects for whom the treatment centers did not respond to requests for information. These patient listings were dropped because they could not be identified as recipients. An additional 833 patients from whom hGH was supplied never actually received the hormone from the physicians. Other patients were dropped from the study because they were treated only for nongrowth-related problems, e.g. osteoporosis (n = 169), because they used only commercial hormone (n = 9), or because their families reported that they never actually used the hormone (n = 30). As a method of verifying the completeness of patient ascertainment, the study investigators asked the Human Growth Foundation, a private organization, to poll their members. The Human Growth Foundation asked its members who had received hGH to identify themselves. In all, 120 members reported receiving NHPP hGH. Ninety-three percent of these (112) had been identified previously by the study investigators.

Table 1. Patients treated

Non-duplicated patient listings, from NHPP applications	8,139
Self-reports from family, no identifiable listing on NHPP application	18
Total possible cohort	*8,157*
Excluded from consideration because:	
Not treated (per center)	833
Not treated (per family)	30
Treated w/commercial only (per center and interview, if done)	9
Patients treated for experimental reason (per center)	169
Cohort of interest to study	*7,116*
Treatment center nonresponse	70
Treatment center unable to provide, identifier/ treatment information	762
Cohort available for analysis	6,284[1]

[1] Included in the cohort are 5 patients who reported in the interview that they received NHPP while the treatment center indicated they received commercial hormone, and 97 patients who reported in the interview that they received commercial hormone while the treatment center indicated they received NHPP.

Tracing hGH Recipients

In some cases, tracing hGH recipients was simple because the recipients were still being treated and their physicians had up-to-date addresses and telephone numbers for them. For many older subjects, however, the process was quite difficult. Many addresses and telephone numbers were outdated and some subjects had married and changed their names.

Because tracing some recipients was difficult, Westat Inc. (the organization that conducted the investigation under contract from the USPHS) used several approaches simultaneously. Telematch, a service which verifies addresses and telephone numbers, was able to verify reported addresses and telephone numbers for just under one third of subjects. Postmasters were contacted to obtain forwarding addresses. Tracing letters were sent to the last known addresses of subjects or their families asking them to call a toll-free number. Other sources were used to obtain up-to-date addresses and telephone numbers including motor vehicle administration records, voter registration records, tax records, school data and criss-cross directories. In some instances, landlords, employers and neighbors were asked to provide tracing information but were not told the reason for our interest.

Because the study planned to conduct a telephone interview with all recipients or their families, it was critical that a current telephone number as well as an address be obtained. Unfortunately, many families had unlisted telephone numbers. In those cases a letter was sent to the family asking them to provide their telephone number. If no response was received, criss-cross directories were used to contact neighbors. If this was unsuccessful, a certified letter was sent to the family, again urging them to contact the investigators.

In all, addresses and telephone numbers were obtained for 5,598 of the 6,284 subjects (89%). Addresses but not telephone numbers were verified for an additional 345, bringing the total traced to 5,943 (95%). There were numerous reasons why the remainder were not traced. Subjects living outside the US were sent a letter asking for basic information on health status and neurological symptoms but were not contacted by telephone. Of the 84 subjects living abroad, only 11 responded. Thirteen families were not contacted at the request of the treating physician, usually for psychological reasons. Finally, the 5 known (US) cases of CJD [7, 8] were not given the standard interview because data were already available on them and the study investigators did not want to risk upsetting family members.

Data Gathering

Interviewing Subjects and Their Families

Among the methods used to gather data for this study, conducting telephone interviews with recipients and their families was the most important. As a general policy parents were contacted first. For adult recipients the parents were asked for information needed to contact the recipient directly. In some cases, parents did not want their grown offspring contacted. Their wishes were respected.

Interviewers asked about subjects' social and demographic status, reasons for GH therapy, other endocrine problems, other medical conditions (particularly neoplasms) and neurological symptoms. Interviewers asked a series of questions relating to symptoms of CJD, e.g. weakness of the arms or legs. Those who reported recent onset (within the preceding 6 months) or worsening of the symptoms and had seen a physician for that complaint were flagged for review by the study investigators. Generally, if the symptoms were suggestive of CJD, medical records were obtained. If the symptoms were equivocal, the subject (or parent) was recontacted in 3–6 months to see if the condition had worsened. Because the course of CJD is rapidly progressive, it was assumed that any subject who had not deteriorated significantly in the interim did not have CJD.

Data were also collected on subjects' height by questionnaire and, where possible, by actual measurement. Subjects were sent a kit containing measuring tape and instructions on how to attach the tape to a door frame and to measure their height. A pre-test of the kit in a small number of subjects who had reached known final adult heights demonstrated that self-measured heights were within 3 cm of heights obtained by medical personnel 87% of the time.

Death Follow-Up

Because of the serious underlying conditions that produced GH deficiency in many of the study subjects, it was not surprising that many of them died. Information on deaths and causes of death were obtained from several sources: physician reports, interview data from families, and the National Death Index. A list of all study subjects was compared with the National Death Index (using date of birth, state of residence and social security number, if available) to identify those who had died and to obtain death certificate numbers. Death certificates were then requested from the state vital statistics offices. Westat obtained death certificates for 225 of the 254 reported deaths (excluding the known CJD deaths). In addition to confirming that death had occurred, the death certificates provided information on causes of death.

Further information on cause of death was obtained from medical records. In all, medical records were obtained on 248 of the 254 non-CJD deaths. All medical records were screened by a study physician. Those who had neurological symptoms which could not be explained by known conditions were referred to a committee of neurologists for independent review.

Pathological information was requested for all deceased subjects; brain pathology slides were borrowed, if available. Neuropathology was available on 93 (37%); 126 (50%) had no brain autopsy and no pathology was available for 33 (13%). The slides were reviewed independently by 2 neuropathologists. Then their reports were submitted along with the medical records to the neurology review committee for final evaluation. These CJD experts rated the likelihood that the subject had CJD.

Other CJD Reporting

In addition to the methods described above, possible CJD cases were also reported directly to study investigators by treating physicians. P.B. and J.F. received several such reports over the course of the study.

Identification of Suspect Lots or Batches of hGH

One of the goals of the study was to determine whether one or more specific lots (of hGH) or batches (of pituitaries used to make hGH) were contaminated with the CJ agent. Study investigators (P.B., J.F.) identified many of the individual lots/batches the patients who developed CJD had received, based on records from the laboratories that produced hGH for the NHPP. The study investigators also obtained data on which lots were

Table 2. Outcomes of attempts to interview GH recipients and their families

Outcome	n	%
Total number in cohort	6,284	
Successfully interviewed	5,240	84
Not interviewed		
Could not be traced	256	4
No telephone number available	346	5
Refused	254	4
Living outside the US	73	1
No one available to interview[1]	18	–
Could not be reached	27	–
Language or other problems	17	–
Interview not attempted (known CJD cases)	5	–
Reached only by mailed questionnaire	48	–

[1] Too ill or incompetent; all family members deceased; physician requested they not be contacted.

sent to which institutions. These data are being used to derive a rough estimate of how many subjects were exposed to each lot and batch. This analysis was based on the number of subjects under therapy at each institution at the time the lots were being distributed.

Current Status

Success at Locating and Interviewing Subjects

Of the 5,943 subjects for whom addresses and/or telephone numbers were available, at least 1 member of 5,240 families had been interviewed by the end of the interview phase of the study in February 1989 (table 2). In all, 5,069 parents were interviewed; 2,020 had children under 18 years of age who were not interviewed according to the protocol; 2,874 had children 18 years of age or older and 173 had deceased children (2 ages were unknown). Although the protocol specified that recipients aged 18 or over were to be interviewed, 1,119 were not interviewed at the parents' request (1,076) or because they could not be reached (43). The remaining 1,755 were successfully interviewed. There were 171 recipients whose parents could not be interviewed but who were interviewed directly. In addition, mail questionnaires were completed by 48 families bringing the total number of subjects for whom interview information was obtained to 5,288 (84.1%)

The reasons for failure to interview subjects are listed in table 2. Inadequate information on the family's place of residence, lack of a telephone number and refusals were the major reasons for failure. Fortunately, Westat staff members were able to obtain some vital status information on some refusers. Of the 254 families who refused to be interviewed, 41 patients were known to be alive and well, and 11 were known to be alive, but had health problems; no information was available on the remaining 202 subjects.

Ongoing Analysis

Data analysis is proceeding in several areas. The death investigation is virtually complete and a summary of causes of death in hGH recipients will soon be prepared. Medical record and pathology reviews for possible cases of CJD are nearly complete. Likewise, the possible cases of leukemia are being investigated. This phase of the investigation should be completed by early 1990.

References

1 Jubelt B, Miller JR: Viral infections; in Rowland LP (ed): Merritt's Textbook of Neurology. Philadelphia, Lea & Febiger, 1989, pp 126–127.
2 Brown P, Gajdusek DC, Gibbs CJ, et al: Potential epidemic of Creutzfeldt-Jakob disease from human growth hormone therapy. New Engl J Med 1985;313:728–731.
3 Brown P: Human growth hormone therapy and Creutzfeldt-Jakob disease: A drama in three acts. Pediatrics 1988;81:85–92.
4 Tintner R, Brown P, Hedley-Whyte T, et al: Neuropathologic verification of Creutzfeldt-Jakob disease in the exhumed American recipient of human pituitary growth hormone: Epidemiologic and pathogenetic implications. Neurology 1986;36:932–936.
5 Koch TK, Berg BO, De Armond SJ, et al: Creutzfeldt-Jakob disease in a young adult with idiopathic hypopituitarism. New Engl J Med 1985;313:731–733.
6 Gibbs CJ, Joy A, Heffner R, et al: Clinical and pathological features and laboratory confirmation of Creutzfeldt-Jakob disease in a recipient of pituitary-derived human growth hormone. New Engl J Med 1985;313:734–738.
7 Marzewski DJ, Towfighi J, Harrington MG, et al: Creutzfeldt-Jakob disease following pituitary-derived human growth hormone therapy: A new American case. Neurology 1988;38:1131–1133.
8 New MI, Brown P, Temeck JW, et al: Preclinical Creutzfeldt-Jakob disease discovered at autopsy in a human growth hormone recipient. Neurology 1988;38:1133–1134.

Dr. J.L. Mills
Prevention Research Program
NICHD/NIH
EPN Building, Room 640
Bethesda, MD 20892 (USA)

Discussion

G. Costin, USA: Dr. Preece's presentation suggested that GH treatment does not have significant side effects on CHO metabolism and insulin action. Results of our studies [Costin et al., Horm Res 1989;31(suppl 1):34, abstract 133] on 9 GH-deficient short patients treated with 0.6–0.7 U/kg/week GH for 1 year indicate a significant and progressive decrease in insulin sensitivity associated with a significant increase in fasting glucose and stimulated glucose and insulin levels at 6 months of therapy. There was a transient rise in the glucose K value at 2–4 weeks of therapy and no significant change in HbA_1 level. I would like to caution of the potential side effects of high-dose GH treatment particularly in non-GH-deficient patients who may also have in some instances a genetic predisposition to diabetes. More sensitive studies than fasting and stimulated glucose and insulin during an OGTT are required to document the initial subtle changes in CHO metabolism and particularly insulin sensitivity.

M.A. Preece, UK: That is a big dose. The data I presented were based on lower-dose studies.

N. Stahnke, FRG: I do not think we really know enough about metabolic actions of long-term GH administration to state that there are no untoward side effects. Especially since the currently used GH doses are definitely higher than those administered formerly: up to 30 U/m^2/week versus 12–14 U/m^2/week. Even after the relatively small GH dose of 18 U/m^2/week we found in Turner patients a significant increase in LDL cholesterol. This may have implications for the development of arteriosclerosis. I think we all should pay more attention to these possible side effects.

M.A. Preece: With respect, that is what I have been saying. At the present time the doses that we have experimented with, you do see side effects. I am not saying we should congratulate ourselves and stop looking.

Z. Laron, Israel: As Dr. Costin mentioned, genetically predisposed subjects to CHO intolerance, such as patients with Turner syndrome who are known to develop CHO intolerance and patients with short stature and a family history of diabetes, should receive special attention and be studied more carefully.

M.A. Preece: I do not think we know the answer to that. It is important to keep that point in mind.

W.M. Teller, FRG: Dr. Mills, could you comment on the evidence of recurrence of brain tumors during hGH replacement therapy? This question has arisen when children have been operated on for brain tumors, craniopharyngiomas, etc. then had GH deficiency and were treated. Should we discuss this with the parents or can we ignore it?

J. Mills, USA: This is a very good question, but a very difficult one to answer. One would need a randomized study in which children were assigned to receive hGH or not without the investigators being influenced by the children's tumor status. Such a study has not been done. Our study cannot address this question because all subjects with tumors had to receive hGH in order to be part of our study.

J. Girard, Switzerland: Would you comment in respect of unwanted effects of hGH – not only on the total dose but on the mode of treatment? It might well be that giving hGH in a 'more physiological' way, i.e. more than once daily, will not only improve the efficacy on height velocity but enhance possible unwanted effects. The closer we go to a 'physiological' regime, the more we might approach a 'mini acromegaly syndrome'.

M.A. Preece: I do not think there is any published data that will answer that. I know there are studies in process at the moment looking at the difference between once to 3 times a day administration of hGH, both in terms of efficacy and in terms of safety, but these studies are still only preliminary. I do not think you will have information for at least 12–24 months.

M. Silink, Australia: I wonder whether you could comment on the state of association between leukemia and hGH deficiency as such, and if you would also comment on the association between leukemia and hGH therapy in the Japanese experience?

J. Mills: In terms of the association between leukemia and hGH deficiency, I have not seen any data that would help me to draw conclusions. What I have seen are reports. Without some kind of denominator to determine what the rate is in a population, I do not think it is a clever idea to speculate as to whether the rate has increased or not. In terms of the Japanese information, it appears that they are turning up more cases in the hGH-treated population. I have not seen any publications of their most recent cases but they have data in terms of how many people have been treated, how many years of follow-up have been included in their analysis and they do seem to find it a fact. However, you would have to ask whether it is the underlying disease or is it the hGH or is it some as yet undetermined factor and they have not yet provided us with an answer to that question.

Horm Res 1990;33:121–127

0301–0163/90/0334–0121$2.75/0

Pharmacological Testing of Growth Hormone Secretion

E.O. Reiter, P.M. Martha, Jr.

Department of Pediatrics, Baystate Medical Center, Tufts University School of Medicine, Springfield, Mass., USA

Key Words. Growth hormone tests · Growth hormone deficiency

Abstract. The laboratory confirmation of growth hormone (GH) deficiency (GHD) has been extensively studied. Multiple stimuli induce GH release, but insulin-induced hypoglycemia usually is considered the 'gold standard'. Seventy-five to 90% of normal children have significant increments of hGH to any single test. Complete and partial syndromes of GHD have been defined, but some patients with a clinical appearance of GHD release hGH during provocative testing. Discordant results on varied tests may occur in the same child. Sequential and simultaneous tests have been attempted with diverse time patterns; testing sequence may significantly affect data interpretation. Persistent problems with GH provocative tests remain: normal data not strictly defined throughout childhood, multiple tests with discordant results, and substantial discrepancies of immunopotency estimates with different radioimmunoassays. Some children with 'normal' hGH increments during provocative tests, despite clinical GHD, may require short-term treatment with hGH to finally establish the diagnosis.

Introduction

In an attempt to substantiate clinical suspicions of growth hormone (GH) deficiency, varied pharmacological tests, as well as so-called 'physiological screening studies' and attempts at evaluation of endogenous secretion of human GH (hGH) have been utilized. We focus in this presentation on pharmacological testing of hGH secretion. Several years ago, Frasier [1] reviewed hGH stimulation tests during childhood. Since that time, no substantial advance has been made in pharmacological assessment of hGH production, whether it be in the types of tests utilized or in our ability to interpret the test data.

With the development of specific and precise radioimmunoassays (RIAs) for GH and the realization of important reciprocal relationships between GH and glucose, the well-established insulin tolerance tests (ITTs) for provoking hGH secretion became available in the early 1960s [2, 3]. This began nearly a 3-decade odyssey in search of the ideal tool to assess hGH sufficiency. Our increasing understanding of the complexity of neuroregulation of hGH production has but amplified the mystery of the journey, recently involving adrenergic and cholinergic agonists or antagonists to examine the modulation of hypothalamic hGH-regulatory factors [4, 5].

Early Studies

The landmark 1968 publication by Kaplan et al. [6] describing changes in levels of hGH during ITTs in a large group of short and normal children remains a standard against which other testing methods are contrasted. In 52 of 53 children with hypopituitarism, many of whom had organic disease, hGH did not rise to a level greater than 2.5 ng/ml. In 1 child with craniopharyngioma, however, the value rose to 5.5 ng/ml. In a group of growth-retarded children, 15–20% had values during an ITT in the range of 3–7 ng/ml, so-called 'blunted' responses, and several had clinically suspected hGH deficiency. Finally, in all normally statured children, most of the short children, but even the few in whom hGH insufficiency was clinically questioned, a response to greater than 7 ng/ml was demonstated during the ITT. These early data predicted accurately the diversity of test

Table 1. Examples of pharmacological GH provocative studies

Insulin-induced hypoglycemia	Pyrogen
Arginine	Glucagon
l-Dopa	Propranolol
Clonidine	GHRH

Table 2. Attempts at overcoming problems with pharmacological hGH provocative testing

Utilization of sequential tests on the same day
Simultaneous administration of several testing agents
Sex steroid priming
Measurements of hGH-dependent peptides (IGF-I, IGF-II)
Administration of GHRH; addition of cholinergic agonists

responses which lay ahead and the interpretive problems inherent to hGH assessment.

The same investigators, realizing limitations to the use of a single test, went on to compare 2 tolerance tests performed separately, the ITT and the arginine tolerance test (ATT) [7]. In a study of 60 short children, 10% had blunted responses, while 28% yielded results which would be interpreted discordantly on 1 of the 2 tolerance tests. So, a total of 38% of patients tested in this group of short children did not yield easily interpretable data, unless one were willing to readily accept biological variability and the inability of so-called 'normal short' children to respond to all tolerance tests. Alternatively, these children may have represented patients with partial hGH deficiency. In a more recent study derived from a large number of patients in the Genentech National Cooperative Study Group, striking variance between peak responses of hGH during ITT or ATT corroborated the older studies.

As the realization has grown that neurotransmitters and neuropeptides act at a variety of brain, hypothalamic and pituitary sites to ultimately modulate GH secretion, endocrinologists have attempted to determine adequacy of GH production, secretion and action by utilization of agents which affect this complex neuroregulatory system at varied sites (table 1). As a result, many different tests of GH secretion have been utilized, with arginine, *l*-dopa and clonidine being the most commonly used, along with the ITT. It is not at all clear that any single test is substantially better than any other [1, 8].

Attempts to Circumvent Problems with hGH Provocative Tests

A number of attempts have been made to overcome the problems inherent in pharmacological studies (table 2). One such early appraoch involved utilization of sequential tests on the same day in an effort to achieve corroborative tests conveniently [9]. While such studies demonstrated apparently adequate hGH release in grouped data, discordant individual results remained in approximately 30% of the subjects. A second pitfall of sequential studies concerns the timing of test agent administration. Several studies have shown that hGH release may be quite dramatically diminished or augmented by modifying the precise temporal relationship for administration of provocative agents [10–12].

Simultaneous provocations of hGH release have also been attempted. Using a combination of arginine and *l*-dopa, Weldon et al. [13] found that certain subgroups of patients with apparent hGH deficiency following individual tests with *l*-dopa or arginine had normal hGH release when the 2 agents were given in combination. This study raised the concern that a combination test could fail to identify children with partial hGH deficiency who might benefit from therapy.

The use of sex steroid priming and the realization that pubertal changes may affect hGH secretion prompted numerous pertinent studies. Estrogen pretreatment augments hGH release in response to a variety of pharmacological tests [1, 8] but its meaning is obscure. We do not routinely use such priming before testing prepubertal patients. Many studies have demonstrated that there are pubertal changes in both endogenous hGH secretion [14, 15], as well as in pharmacologially provoked hGH release [1, 8]. Gourmelen et al. [16] demonstrated that, in a group of prepubertal patients with constitutional growth delay, 23% had pharmacologically provoked hGH release below the normal range and only slightly above the hGH-deficient range. When these patients with constitutional growth delay finally developed pubertal changes or received sex steroid treatment, hGH release normalized. Those data were interpreted as evidence for transient partial hGH deficiency in these subjects.

To obviate the use of hGH provocative tests, many investigators have used measurements of peptides whose production is largely regulated by hGH. This seemingly reasonable use of insulin-like growth factor I (IGF-I) determinations, however, has been fraught with difficulty, not the least of which is due to the presence of IGF-binding proteins which may lead to erroneous IGF potency estimates [17]. IGF levels are also low in many patients with diverse disorders such as hypothyroidism, malnourishment, renal disease or diabetes [18–21]. Normal and hGH-deficient children have levels of IGF-I which overlap almost completely in very young children [22]. Both normal and low levels of IGF-I are present in short children with diverse diagnoses, such as constitutional growth delay or genetic short stature [23, 24]. Peak provoked hGH levels during pharmacological tests and IGF-I values do not correlate well, with discordance rates of nearly 30% [25]. Additionally, baseline or stimulated levels of IGF-I have not been predictive of subsequent growth responses to hGH treatment, especially if prior acid separation of IGF-I from its binding proteins was not performed [26].

However, studies by Rosenfeld et al. [27] indicate that concomitant measurements of both IGF-I and IGF-II substantially increase the value of IGF measurements in assessment of short stature. In less than 1% of normal children, both IGF-I and IGF-II levels were low, whereas normal values were detected in only 4% of GH-deficient children. Further, among their normal short children, only 11% had low levels of IGF-I and IGF-II, perhaps helping to dissect that group from the group with true hGH deficiency. Additionally, Blum et al. [28] have recently used measurement of IGF-binding protein-3 to identify hGH-deficient patients with a high degree of precision and specificity.

Finally, GH-releasing hormone (GHRH) administration appeared to provide a potentially useful tool for assessing readily releasable pituitary hGH, presumably reflecting degrees of previous exposure to endogenous GHRH. Unfortunately, its usefulness has been limited due to the extreme variability of somatotrope responsivity, even in normal subjects. Further, peak hGH levels after GHRH overlap substantially between short normal children and those with partial and severe hGH deficiency [29, 30]. The addition of a cholinergic agonist, pyridostigmine, putatively to suppress somatostatin release, may yet salvage this testing system. Ghigo et al. [31] recently showed that combined use of pyridostigmine and GHRH will increase hGH release over that induced by GHRH alone and may help separate hGH-deficient patients from those with constitutional growth delay or familial short stature. Further studies are required to substantiate these data. Finally, the temporal relationship of GHRH administration to another provocative test, for example, an ITT or clonidine, may significantly alter hGH release [10, 11].

Relationship of Endogenous Measurements of hGH Secretion to Pharmacological Testing

In several recent studies [32, 33], workers have suggested great difficulty in separating mean 24-hour hGH levels in normal prepubertal children from those of short prepubertal children. Further, in one study [32], only 57% of patients who had abnormal hGH release following several pharmacological provocative tets, had low mean 24-hour hGH levels; no subjects with normal provocative test results also had low endogenous levels. In contrast, other investigators [34, 35] claimed that studies of endogenous hGH secretion provide clinically useful data, as well as important physiological information. However, because many centers cannot conveniently perform the laborious 24-hour testing, significant effort is being directed toward validating 24-hour urinary hGH measurements as an integrated reflection of hGH production. At this time, we do not routinely use frequent measurements of hGH during a 12- or 24-hour sampling period in the evaluation of short children.

Variability of GH Immunopotency Estimates

Recent studies have raised the issue of variability of the measurement of GH in the multiple RIA systems available [36–38]. As much as 2- to 4-fold discrepancies have been noted when different assay systems have been utilized to measure GH immunopotency. The discordance of measurements with monoclonal and polyclonal antisera is largely related to matrix effects in the assay systems, but also due to molecular heterogeneity of circulating GH species, differences in GH standards, purity of GH tracer and basic method of measurement (competitive binding vs. sandwich assay). Additionally, al though it is likely that the high-affinity GH antisera used in most immunoassay systems would obviate problems related to the presence of GH-binding proteins in the serum, such effects have not been systemically evaluated.

In addition to the variability of immunopotency estimates, laboratory diagnostc criteria for hGH deficiency have generally loosened over the years as the supply of hGH has risen (fig. 1). The broadening of the arithmetic definition of hGH deficiency from levels of less than 2.5 ng/ml in the 1960s to less than 10 ng/ml at present and perhaps higher in the future may or may not prove to be a clinically wise evolution.

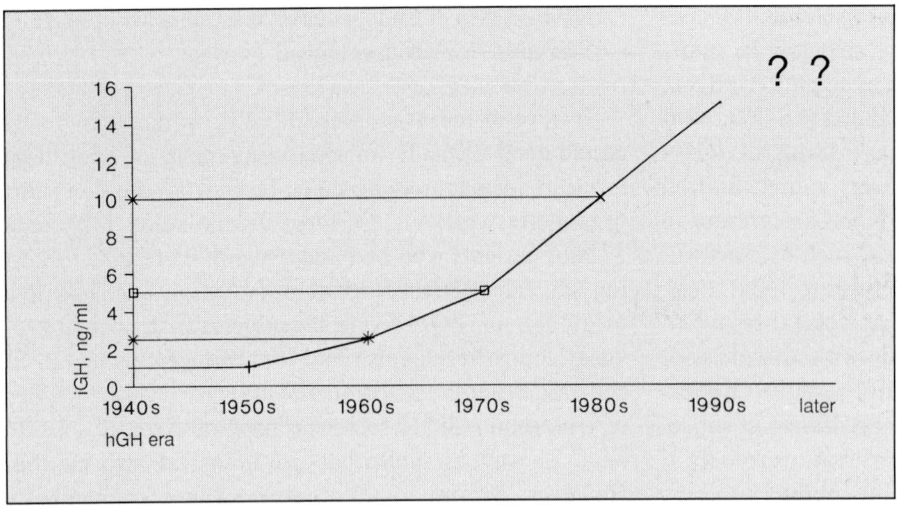

Fig. 1. Schematic representation of the hGH levels used to diagnose hGH deficiency with provocative testing during the area of hGH availability.

Table 3. Relationship of peak provoked hGH levels to growth induced by 1 year of hGH therapy

Group	Age years	Height (SD) SDS	Growth velocity, cm/year	
			pre-hGH	hGH treatment
Peak hGH <3.5 ng/ml	9.9±5.2 (n = 180)	−3.5±1.3 (n = 169)	4.2±1.9 (n = 41)	9.5±3.6 n = 74)
Peak hGH >7.5 ng/ml	11.1±3.5 (n = 251)	−2.9±0.8 (n = 247)	4.5±1.9 (n = 72)	8.1±1.9 (n = 81)

The two tests used were ITT and ATT. Data were obtained from the Genetech National Cooperative Growth Study and are expressed as mean SD.

Relationship of Provoked hGH Levels to Subsequent Growth Responses

The relationship of peak hGH levels during 2 provocative tests (ITT and ATT) to incremental growth responses during 1 year of exogenous hGH treatment has been examined (table 3; Genentech National Cooperative Growth Study). In those individuals in whom the diagnosis of severe hGH deficiency would be made (peak hGH on 2 tests less than 3.5 ng/ml), the 12-months growth velocity was 9.5 ± 3.6 (SD) cm; in those individuals in whom the diagnosis of hGH insufficiency generally would not be entertained (peak hGH on both tests greater than 7.5 ng/ml), growth velocity over that 12-month period was 8.1 ± 1.9 cm. The latter group also had a generous mean increment of velocity of nearly 4 cm/year.

Although the lowest pharmacological test responders with 'classic' hGH deficiency had more of a growth increment, we cannot argue that peak-provoked hGH release is necessarily predictive of subsequent growth induced by exogenous hGH treatment. Juxtaposed to these data, however, several investigators have demonstrated apparently much clearer inverse relationships between endogenous hGH-secretory data and treatment-induced growth increments [39, 40]. Better yet, perhaps, auxologic data, such as pretreatment height or height velocity, may have a predictive value as adequate as any of the testing systems.

Conclusions

Difficulties continue with the interpretation of hGH testing, even after 3 decades. A definition of truly normal data on an age- and sex-related basis for provocative testing is lacking. The explanation for the discordance of peak-provoked hGH results for the many tests is ascribed to biological variability or different neuroregulatory abnormalities although one cannot forget potential temporal effects in the sequential studies. The problem of variable immunopotency estimates with diverse commercial assays and the general loosening of diagnostic criteria may be an important issue, especially when third-party payers are involved. Finally, there is a questionable relationship of provoked hGH levels to subsequent growth responses to exogenous hGH treatment.

To close the circle and bring us back to the important paper of Kaplan et al. [6], an important clinically relevant conclusion was drawn – not dissimilar to those in very recent papers:

'In some patients who have an apparently normal serum immunoreactive growth hormone response to a provocative test, despite physical stigmata of hypopituitarism, short-term therapy with growth hormone may be necessary to establish a diagnosis of growth hormone insufficiency.'

This 1968 statement reflects our present opinion that results of hGH pharmacological testing cannot uniquely or absolutely determine the decision to initiate hGH therapy in an individual child. Critical scrutiny of laboratory studies and appropriate assessment of auxologic data should lead to cogent, but surely reversible clinical judgments, possibly utilizing predefined growth responses to a predefined period of hGH therapy to confirm or reject that clinical decision. We presently use pharmacological testing of hGH secretion in that clinical context, but await the results of longer-term trials with hGH therapy in patients who do not meet older and more restrictive definitions of hGH insufficiency before routinely recommending hGH therapy to such individuals.

Acknowledgments

We are grateful to the Genentech National Cooperative Growth Study Group for providing data for this presentation and for the expert secretarial assistance of Sarah Harris.

References

1 Frasier SD: A review of growth hormone stimulation tests in children. Pediatrics 1974;53:929–937.
2 Glick SM, Roth J, Yalow RS, et al: Immunoassay of human growth hormone in plasma. Nature 1963;199:784–787.
3 Roth J, Glick SM, Yalow RS, et al: Hypoglycemia: a potent stimulus to secretion of growth hormone. Science 1963;140: 987–990.
4 Bercu BB, Diamond FB: A determinant of stature: Regulation of growth hormone secretion. Adv Pediatr 1986;33:331–380.
5 Frohman LA, Jansson JO: Growth hormone-releasing hormone. Endocr Rev 1986;7:223–253.
6 Kaplan SL, Abrams CAL, Bell JJ, et al: Growth and growth hormone. I. Changes in serum level of growth hormone following hypoglycemia in 134 children with growth retardation. Pediatr Res 1968;2:43–63.
7 Youlton R, Kaplan SL, Grumbach MM: Growth and growth hormone. IV. Limitations of the growth hormone response to insulin and arginine and of the immunoreactive insulin response to arginine in the assessment of growth hormone deficiency in children. Pediatrics 1969;43:989–1004.
8 Chernausek SD: Laboratory diagnosis of growth disorders; in Hintz RL, Rosenfeld RG (eds): Growth Abnormalities. Contemporary Issues in Endocrinology and Metabolism. New York, Churchill Livingstone, 1987, vol 4, pp 231–254.
9 Penny R, Blizzard RM, Davis WT: Sequential arginine and insulin tolerance tests on the same day. J Clin Endocrinol Metab 1969;29:1499–1501.
10 Reiter EO, Morris AH, Biggs DE: Modulatin of GHRH-induced growth hormone release by an alpha-adrenergic agonist and hypoglycemia. J Pediatr Endocrinol 1988;3:21–25.
11 Shibasaki T, Hotta M, Masuda A, et al: Plasma GH responses to GHRH and insulin-induced hypoglycemia in men. J Clin Endocrinol Metab 1985;60:1265–1267.
12 Woolf PD, Lantigua R, Lee LA: Dopamine inhibition of stimulated growth hormone secretion: Evidence for dopaminergic modulation of insulin- and *l*-dopa-induced growth hormone secretion in man. J Clin Endocrinol Metab 1979;49:326–330.
13 Weldon VV, Gupta SK, Klingensmith G, et al: Evaluation of growth hormone release in children using arginine and *l*-dopa in combination. J Pediatr 1975;87:540–544.
14 Finkelstein JW, Roffwarg HP, Boyar RM, et al: Age-related change in the twenty-four hour spontaneous secretion of growth hormone. J Clin Endocrinol Metab 1972;25:665–670.
15 Martha PM, Rogol AD, Veldhuis JD, et al: Alteration in the puslatile properties of circulating growth hormone concentrations during puberty in boys. J Clin Endocrinol Metab 1989;69: 563–570.
16 Gourmelen M, Pham-Huu-Trung MT, Girard F: Transient partial : deficiency in prepubertal children with delay of growth. Pediatr Res 1979;13:221–224.
17 Zapf J, Walter H, Froesch ER: Radioimmunological determination of insulin-like growth factors I and II in normal subjects and in patients with growth disorders and extra pancreatic tumor hypoglycemia. J Clin Invest 1981;68:1321–1330.
18 Clemmons DR, Klibanski A, Underwood LE, et al: Reduction of plasma immunoreactive somatomedin-C during fasting in humans. J Clin Endocrinol Metab 1981;53:1247–1250.
19 Baxter RAC, Brown AS, Turtle JR: Radioimmunoassay for somatomedin-C: Comparison wth radioreceptor assay in patients with growth disorders, hypothyroidism and renal failure. Clin Chem 1982,28:488–495.
20 Soliman AT, Hassan AE, Aref MK, et al: Serum insulin-like growth factors I and II concentrations and growth hormone and insulin responses to arginine infusion in children with protein-energy malnutrition before and after nutritional rehabilitation. Pediatr Res 1986;20:1122–1130.
21 Tamborlane WV, Hintz RL, Bergman M, et al: Insulin infusion pump treatment of diabetes: influence of improved metabolic control on plasma somatomedin levels. N Engl J Med 1981;305: 303–307.
22 Bala RM, Lopatka J, Leung A, et al: Serum immunoreactive somatomedin-C levels in normal adults, pregnant women at term, children at various ages and children with constitutional delayed growth. J Clin Endocrinol Metab 1981;52:508–512.
23 Van Vliet G, Styne DM, Kaplan SL, et al: Growth hormone treatment for short stature. N Engl J Med 1983;309:1016–1022.
24 Rubin KR, Lichtenfels JM, Ratzan SK, et al: Relationship of somatomedin-C concentration to bone age in boys with constitutional delay of growth. Am J Dis Child 1986;19:975–980.
25 Reiter EO, Lovinger RD: The use of commercially available

somatomedin-C radioimmunoassay in patients with disorders of growth. J Pediatr 1981;99:720–724.
26 Sherman BM, Frane J, Hintz RL, et al: IGF-I measurement in growth hormone deficiency: The clinical relevance and dependence on method (abstract 511). 98th Annu Meet Am Pediatr Soc, Washington, May 1988. Pediatr Res 198;23:286A.
27 Rosenfeld RG, Wilson DM, Lee PDK, et al: Insulin-like growth factors I and II in evaluation of growth retardation. J Pediatr 1986;109:428–433.
28 Blum WF, Ranke MB, Kietzmann K, et al: A specific radioimmunoassay for growth hormone (GH)-dependent somatomedin-binding protein: Its use for diagnosis of GH deficiency. J Clin Endocrinol Metab 1990;70:1292–1298.
29 Schriock EA, Lustig RH, Rosenthal SM, et al: Effect of growth hormone (GH)-releasing hormone (RGH) on plasma GH in relation to magnitude and duration of GH deficiency in 26 children and adults with isolated GH deficiencies. J Clin Endocrinol Metab 1984;58:1043–1049.
30 Albini CH, Reiter EO, Mills BJ, et al: Diagnostic value of the growth hormone-releasing factor stimulation test. Clin Pharmacol Ther 1988;43:595–700.
31 Ghigo E, Mazza E, Imperiale E, et al: Growth hormone responses to pyridostigmine in normal adults and in normal and short children. Clin Endocrinol 1987;27:669–673.
32 Rose SR, Ross JL, Uriarte M, et al: The advantage of measuring stimulated as compared with spontaneous growth hormone levels in the diagnosis of growth hormone deficiency. N Engl J Med 1988;319:201–207.
33 Lin T-H, Kirkland RJ, Sherman BM, et al: Growth hormone testing in short children and their response to growth hormone therapy. J Pediatr 1989;115:57–63.
34 Hindmarsh P, Smith PJ, Brook CGD, et al: The relationship between height velocity and growth hormone secretion in short prepubertal children. Clin Endocrinol 1987;27:581–591.
35 Albertsson-Wikland K, Rosberg S: Analyses of 24-hour growth hormone profiles in children: Relation to growth. J Clin Endocrinol Metab 1988;67:493–500.
36 Blethen SL, Chasalow FI: Use of a two-site immunoradiometric assay for growth hormone (GH). J Clin Endocrinol Metab 1983; 57:1031–1035.
37 Reiter EO, Moris AH, MacGillivray MH, et al: Variable estimates of serum growth hormone concentrations by different radioassay systems. J Clin Endocrinol Metab 1988;66:68–71.
38 Celniker AC, Chen AB, Wert RM, et al: Variability in the quantitation of circulating growth hormone using commercial immunoassayxs. J Clin Endocrinol Metab 1989;68:469–476.
39 Albersson-Wikland K: Growth hormone treatment in short children. Acta Paediatr Scand [suppl] 1986;325:64–70.
40 Hindmarsh PC, Smith PJ, Pringle PJ, et al: The relationship between the response to growth hormone therapy and pre-treatment growth hormone secretory status. Clin Endocrinol 1988; 28:559–563.

Dr. E.O. Reiter
Department of Pediatrics
Baystate Medical Center
Tufts University School of Medicine
Springfield, MA 01199 (USA)

Discussion

I. Richter, FRG: I would like to address the problem of tissue sensitivity. The difficulties and doubts of all well-known tests can be avoided only by using the ^{15}N-glycine test [Richter et al., J Clin Endocrinol Metab 1987;65:74–77]. ^{15}N-glycine is administered orally, followed by timed urine collections. I am surprised that nobody has mentioned this test in spite of the fact that Prof. Prader introduced the nitrogen balance test many years ago. We have much experience with ^{15}N-glycine which is a noninvasive test. Two tests are necessary, one before and one after administration of hGH. From data of total N input, ^{15}N dose, total urinary N excretion, urinary ^{15}N output and urine volume, we can determine rates of protein syntesis, protein breakdown, net protein gain and protein turnover using a 3-pool method [Sprinson and Rittenberg, J Biol Chem 1949;180:715]. We have also tested 10 children who were treated by recombinant hGH Saizen (Sereno, Italy), 1.7 U/m^2/day for 8 days. the protein synthesis rate increased from 2.35 ± 0.67 to 4.17 ± 1.35 g/kg/day with a net protein gain from 0.56 ± 0.45 to 1.10 ± 0.43 g/kg/day. There was also a 50% reduction in the urinary nitrogen excretion. We think that the ^{15}N-glycine test has the advantage of not only differentiating between different forms of hGH deficiency, but can also help determine whether hGH therapy will be successful.

E. Reiter, USA: I would agree that an assessment of nitrogen retention would be an important indicator of response to hGH. Nonetheless, in contrast to the data reported by the questioner, others do not have information which is quite as clear-cut.

Z. Laron, Israel: Dr. Reiter, you mentioned that age- and sex-related data for hGH provocative tests are missing. Our group has published such data for insulin hypoglycemia and arginine [Josefsberg et al., Diabetes Metab 1978;4:103–107; Josefsberg et al., Pedatr Adolesc Endocrinol. Basel, Karger, 1983, vol 12, pp 66–74].

These data involving 477 tests in 389 children and adolescents clearly revealed that during peak puberty there was a greater response of hGH than during pre- or early puberty, more accentuated in males. In addition, I wish to say that we found the GHRH test useful in differentiating between hGH deficiency of pituitary or hypothalamic origin. [Keret et al., Israel J Med Sci 1988;24:207–211; Laron and Bauman, Eur J Pediatr 1986;145:6–9]. Having an experience of thousands of provocative tests over years, we found that sex hormone priming is very useful in discriminating true from false hGH deficiency in peripubertal years [Laron et al., Clin Endocrinol 1972;1:91–97] and that when interpreting the normalcy of a hGH response one has to consider, in addition to the pubertal rating, at what timing after the administration the respective drug exerts its maximal effect and consider the normal range for that stimulus [Laron et al., Pediatr Adolesc Endocrinol. Basel, Karger, 1987, vol 16, pp 13–29]. With some agents such as clonidine, the response is dose-dependent [Laron et al., Acta Paediatr Scand 1982; 71:847–848].

J. Girard, Switzerland: There is much to be said for urinary hGH. The assay is easily performed on untreated urine. Although we do not claim to have true normal values, we set up percentiles for age groups for 24 h and first morning urine. Nobody would be expected to distinguish with urine output the continuum from 'normal' to insufficiency. There is no definition of 'normal hGH secretion' for a given height velocity, age, etc. Therapy with hGH is based on a 'greater supply'. The adequacy or excess of such a therapy can easily be assessed by measuring hGH excreted in urine under treatment. Furthermore, doses might be adapted for age or puberty rating. In normal children, urine hGH increases in puberty and shows a clear sex difference.

E. Reiter: I would agree that the assessment of hGH urinary excretion has much to say for it. Nonetheless, it is a very small percentage of produced hGH and is subjected to the difficulties of being associated with abnormalities in glomerular function, as hGH excretion increases in parallel to albumin excretion.

M. Vandweghe, Belgium: Dr. Albertsson-Wikland, you showed in one of your slides that in normal children going through puberty there was a clear-cut rise in 24-hour hGH secretion in stages 3 and 4 of puberty. Did you find any differences between girls and boys in this respect?

K. Albertsson-Wikland, Sweden: The data we have support that the increase in the boys comes later.

Z. Laron: Dr. Albertsson-Wikland, do you consider the 24-hour secretory pattern a test to be advocated for diagnosis or only as a clinical research tool? What do you think about exercise as a screening test and in addition night only versus 24-hour secretory pattern?

K. Albertsson-Wikland: I think that today the 24-hour profile is most helpful for obtaining further information on the physiology but it will not give clear-cut short-term information. It is posible to look at the night period alone since most children secrete low amounts of hGH spontaneously.

Z. Laron: Dr. Reiter illustrated that a growth velocity of 4 cm or more is a normal growth velocity. However, many children below the 3rd percentile have this growth velocity and a marginal hGH reserve which, depending upon the test assay, can be considered normal or abnormal. What is the importance of the growth velocity regarding the decision whether children should be treated or not with hGH.

E. Reiter: I would agree that assessment of growth velocity is exceedingly important in the determination of the possible diagnosis of hGH insufficiency. I would agree with Dr. Prader that 'growth velocity is as important as provocative tests'.

K. Albertsson-Wikland: Most of us have problems with endocrine testing for hGH insufficiency and we have hoped that auxology would help us but I have a feeling that this is not always simple.

E. Reiter: A comment about auxology is that, in an era when in many of our countries we are not in control of who actually receives hGH, the organization which pays for it whether it be the government or life insurance companies, is not always willing to cover the bill because of the notion that auxological data are not completely usable even though many of us would suggest that they are.

A. Prader, Switzerland: I would like to confirm what Dr. Albertsson-Wikland mentioned, and that is that we tend to believe that growth velocity is hard data; but it is not, it is full of errors and I think it is at least as weak as provocation tests or any other tests. We should not forget this especially if it is calculated from short periods of time.

D.L. Donaldson, USA: How extensively should a child who is to be started on hGH therapy be tested? The reason I ask is that our experience has been that stimulated values underestimate spontaneous values whereas a value on just 1 night of spontaneous testing is quite reliable for predicting a second night at least in the low range, although there is a wide variation at higher values. Should these patients be extensively or lightly studied?

E. Reiter: I believe that a short child who is growing poorly should have standard hGH provocative tests to confirm the diagnosis of hGH insufficiency. Should that be done I would feel that no further endocrinologic assessment ought to be undertaken. In the US, such provocative test data remain important for third-party payers.

Z. Zadik, Israel: The reproducibility of the integrated concentration of hGH calculated for 40 tests was $r = 0.859$, $p < 0.01$. The reproducibility of repeating the test in the same subject was $r = 0.472$, $p < 0.01$ ($n = 83$). The best correlation to height velocity increment in a 3-year clinical trial was the pretreatment height velocity.

E. Reiter: I would agree with Dr. Zadik that assessment of reproducibility of pharmacological and physiological tests is important. He demonstrates an extremely high correlation coefficient of the integrated concentration of hGH though this does contrast to data reported by others, especially in the US. It is also my understanding that the pretreatment height velocity was not as clearly related to posttherapy height velocities in the Genentech collaborative trial.

J. Grunt, USA: In light of the fact that over a period of 30 years 30–50% of our very short patients with so-called 'normal hGH' grew very well when treated with hGH, why do we test? Do we test for clinical purposes, for the insurance companies or for pure research purposes? If your answer is what I hope it will be, do you visualize a time where in reality from a clinical point of view, we may not test?

E. Reiter: I believe the reason that we test patients who are to be treated with hGH is to confirm the clinical diagnosis of hGH insufficiency. The fact that there are other patients who do not have hGH insufficiency, but who do grow quite well initially with hGH treatment is well known. These are still research patients and, as such, should have as much characterization of their hGH secretion as possible.

J. Marks, USA: We have looked at this issue from the point of view of the individual child, which as humanists is appropriate. However, we must relate also to the limited resources, particularly financial, to its use. I suspect, Dr. Reiter, that you have been asked by the Commonwealth of Massachusetts to help set standards for prescribing treatment for medically indigent children. I am not asking you to answer today what standards you would set for an insurance company or a state government, yet it is something that every LWPES and ESPE member should think about.

E. Reiter: I agree that the issues of cost of hGH are very important. When placed in the context of the constraints of the health care environment of a state such as Massachusetts, I would agree that limitation of inappropriate utilization of hGH is quite important.

Horm Res 1990;33:128–136

0301–0163/90/0334–0128$2.75/0

Non-Conventional Use of Growth Hormone: European Experience

F. Darendeliler, P.C. Hindmarsh, C.G.D. Brook

Endocrine Unit, The Middlesex Hospital, London, UK

Key Words. Growth hormone · Non-conventional use · European experience · Growth hormone treatment · Short stature

Abstract. Results from several centres in Europe using biosynthetic human growth hormone (hGH) for the promotion of growth in a variety of conditions other than classical hGH deficiency were evaluated. Significant increments in growth rates were achieved by daily administration of hGH in doses appropriate for body size without disproportionate skeletal advances in short normals, Turner syndrome, low birth weight, skeletal dysplasia, central precocious puberty (reared with gonadotrophin-releasing hormone analogue) and renal failure.

Introduction

The limited supply of pituitary-derived human growth hormone (hGH) previously available was used mainly in hGH-deficient patients defined by a series of more or less strict criteria. In a few studies on the use of hGH in conditions other than hGH deficiency, such as Turner syndrome and low birth weight (LBW), a failure of response was reported, although the responses of some of the patients were actually not too bad [1]. The poor responses can now be seen to be attributable to the age of some of the patients treated and to the manner in which hGH was administered.

It is now clear that hGH secretion increases with age [2, 3], especially in puberty [4, 5], so that a fixed-dose regimen in childhood has no scientific basis. We have also learned that giving hGH by daily injection is more effective in promoting growth [6].

Now that an abundant supply of biosynthetic GH (r-hGH) is available, clinicians are able to re-evaluate the indications for the use of this substance in a wide variety of conditions other than classical hGH deficiency.

Short Normals

There has been little agreement about making this diagnosis but the hallmark of normality is growth at a normal rate. Trials of therapy in children growing slowly, whatever their hGH-secretory response to provocative stimuli, do not address the crucial issue for which an answer is demanded by society which is whether adult stature of normal individuals can be increased by hGH and what are the complications of the treatment.

The principal determinant of the rate at which children grow is the amount of hGH they secrete [2, 7, 8] (fig. 1). Hence, short children are short and remain short because they grow continuously less quickly and secrete less hGH than the taller ones. Since there is a continuum of both hGH secretion and growth rate, it is difficult to define a cut-off point for either which would constitute an indication for treatment.

Pharmacological tests of hGH secretion should not be taken solely as a basis of normality because they do not necessarily reflect physiological secretion [9] and they also show a continuum [10]. The normal range in children has not been clearly defined and the reproducibility of the tests in adults is poor. Thus, there cannot be a clear point for any measurement of hGH secretion which can be defined as sufficient or insufficient. Claims that one test is superior to another are based purely on the assumption that the 'gold standard' is perfect [11]. This simply is not the case.

A child growing along the 3rd centile on a distance chart has a height velocity standard deviation score (SDS) of −0.8 and we regard this as the cut-off of normality. However, since the rate at which children grow depends on the amount of hGH they secrete, it becomes

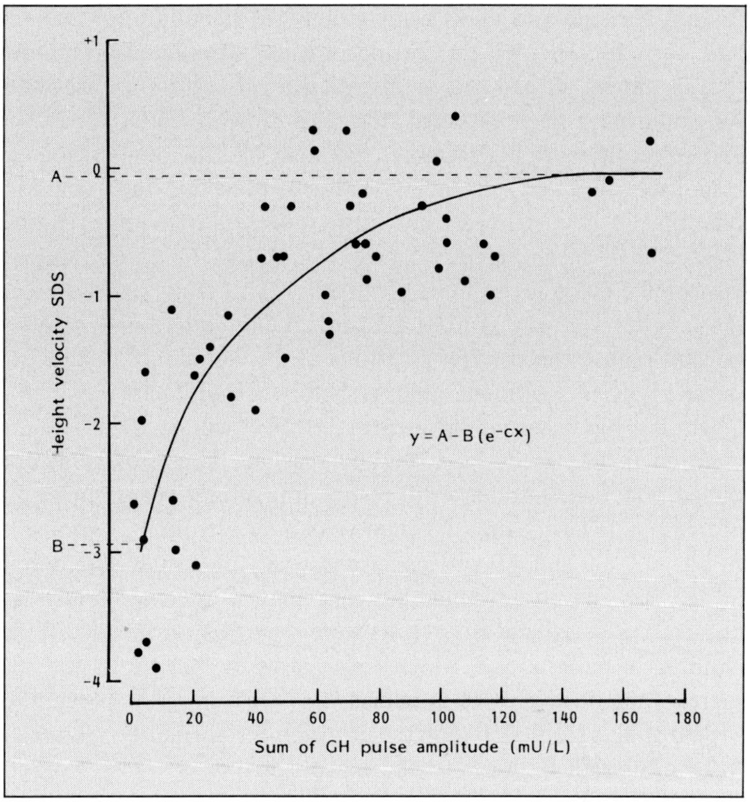

Fig. 1. Relationship between height velocity SDS for chronological age and sum of hGH pulse amplitudes [7].

evident that all children will grow when given hGH. The magnitude of the response will be defined by pre-treatment height velocity [1, 12, 13], hGH-secretory status [14] and dose of hGH administered [12, 15].

We have demonstrated that 3 distinct dose-response curves exist for 3 different height velocity groups [16] (fig. 2). Hence, the faster a child is growing before treatment the bigger is the dose needed to bring about the same change in height velocity. In evaluating the growth response of children with different conditions, these points have to be taken into account.

The European experience on the treatment of short normals is relatively little. Most of the patients treated as short normals have actually not been normal with respect to their growth rate and have demonstrated different degrees of impaired hGH secretion.

Data on the 4 subjects diagnosed as 'small/delay' in Tanner et al.'s study [1] were not conclusive. The situation in 3 non-responders was complicated by psychosocial problems, the start of puberty and a severely compromised parental height (137.7 cm). The Human Growth Hormone Committee's trial in short slowly growing children, which was terminated in 1985 due to the withdrawal of pituitary hGH from use, included 11 children with normal response to pharmacological tests [17]. Only 5 were growing normally, 2 of whom did not respond to therapy. Results of only a 6-month duration of treatment and the giving of hGH in a dose of 4 U thrice weekly to an age range between 2.6 and 12.7 years bring difficulties to the interpretation of results.

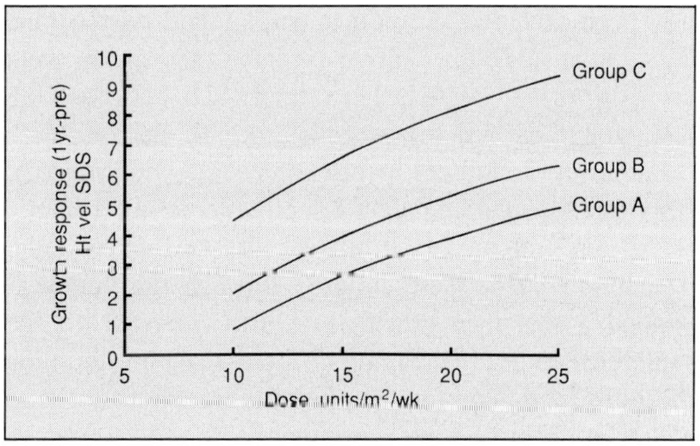

Fig. 2. Dose-response curves for hGH [11]. Group A were children with a pre-treatment height velocity (Ht vel) SDS for chronological age > -0.8, group B had one of -0.8 to -2.0, and group C, < -2.0.

In Albertsson-Wikland's study group [18] all children had normal responses to pharmacological stimuli, although almost all had low or impaired hGH secretion. As would have been expected, the ones growing better and secreting more hGH before treatment had the least response to a given dose of hGH (fig. 3; note that the relationship is curvilinear, not linear). 40 children increased their mean height velocity from 4.6 to 7.5 cm/year on 0.1 U/kg/day of hGH given daily. This study on the long-term effects of hGH [18] brings out some important points: the younger a child is at the commencement of hGH treatment, that is before a significant height loss has occurred, the greater the effect will be. 24 children in this study have already reached their predicted final height and are still responding after 4 years of treatment. The change in height velocity was from 4.2 to 8.1, 6.7, 6.0 and 4.9 cm/year, respectively, in consecutive years and no unwanted advancement of bone age occurred. Less satisfactory results were recorded for the children of pubertal age, which emphasises the need to increase the dose of hGH in puberty.

Wit et al.'s study [19] took the criterion for normality to be a 'normal response to pharmacological tests' and included patients with a height velocity SDS below −2.0. Most must have had impaired hGH secretion. 7 subjects had been of LBW and it would have been better if they had been included in another trial; 1 had dysmorphic features in addition. Only 6 of this study group were growing normally with height velocity SDS between +0.8 and −0.8. Such a heterogeneous group of children with different pre-treatment features showed a wide range of response, the median height velocity increasing from 4.4 to 7.3 cm/year on 14 U/m^2/week of hGH given daily. The follow-up of these children through the 2nd year yielded a sustained growth rate (6.1 cm/year) on the same dose and a higher rate (7.3 cm/year) on doubling the dose [pers. commun.]. The authors could not find any relationship between pre-treatment hGH-secretory status and height velocity SDS for bone age on treatment. However, when height velocity SDS for chronological age was calculated, a tendency for a curvilinear relation became evident, even with such small numbers. We do not understand the relevance of comparing height velocity with bone age. Bozzola et al.'s study [20] (6-month results) yielded a smaller response in the children secreting more hGH before treatment than the ones secreting less hGH on the same dose of hGH, as would have been expected.

The trial carried out at the Middlesex Hospital has included only short children growing normally with normal responses to pharmacological stimuli and normal 24-hour hGH profiles [15]. After 1 years results (n = 16) showed a height velocity SDS increment from −0.44 to +2.20 ($p < 0.001$) on 12–20 U/m^2/week of hGH, whereas the change in the control group (n = 10) was not significant (fig. 4). There was a significant curvilinear relation between growth response and dose (fig. 2, 5). The results of the longer-term follow-up of these children are shown in figure 6. Height velocity SDS was significantly higher than the pre-treatment values at the end of the 3rd year as well as height SDS for chronological age which showed a progressive increase in the same period. The net change in height SDS for bone age was slightly but not significantly higher than in the controls. However, height prediction according to the Tanner-Whitehouse method yielded an increase in final height of 6.8 cm in the boys and 4.2 cm in the girls. No adverse effects of treatment were encountered.

Trials on short normal children are also being carried out in other centres in Europe but the duration of treatment is too short to be conclusive [Vanderschueren-Lodeweyckx, Zachmann, in collaboration with French workers, pers. commun.].

Turner Syndrome

Several early studies of hGH treatment on the growth rate of girls with Turner syndrome yielded moderate to poor results. Inappropriate doses of hGH for body size were given in a fixed-dose regimen. Pre-treatment hGH-secretory status was not taken into account in the evaluation of response.

Tanner et al.'s study [1] included 5 Turner girls, the height velocities of whom are shown in figure 7. It is apparent that the increment in height velocity was associated with hGH dose. It has also become obvious that the older girls would have needed oestrogen as well as hGH, because hGH is not the major determinant of the pubertal component of growth. At this older age the velocity of normal children on the childhood curve is very slow. 8 girls in Stahnke's study [21], who started on hGH at a mean age of 16.0 years and in much lower doses than in the former study, may not have responded for similar reasons. The Human Growth Hormone Committee's fixed-dose trial on 30 patients with Turner syndrome was stopped in 1985 and the short-term results over 6 months were not conclusive [17]. There are single case reports on girls (with or without hGH deficiency) treated with hGH [22–24], the results of which are controversial.

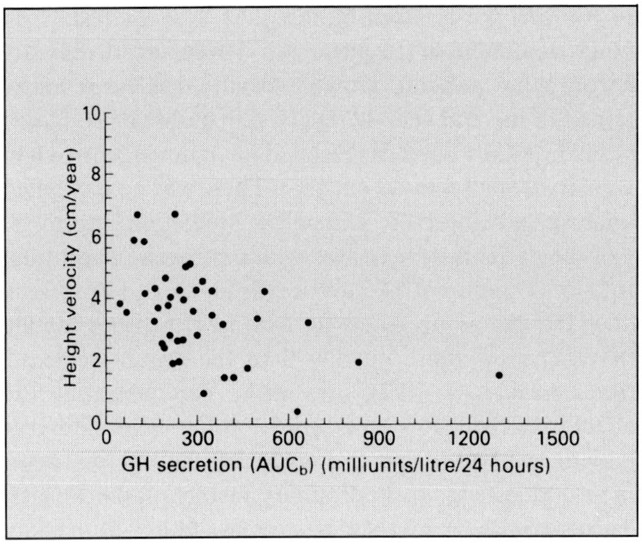

3

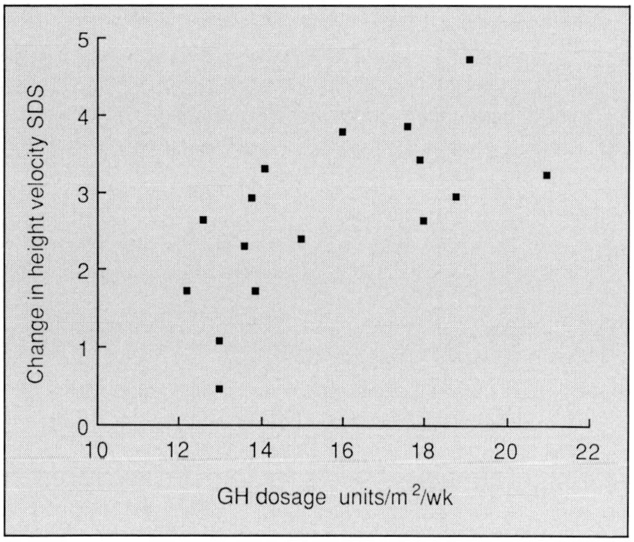

5

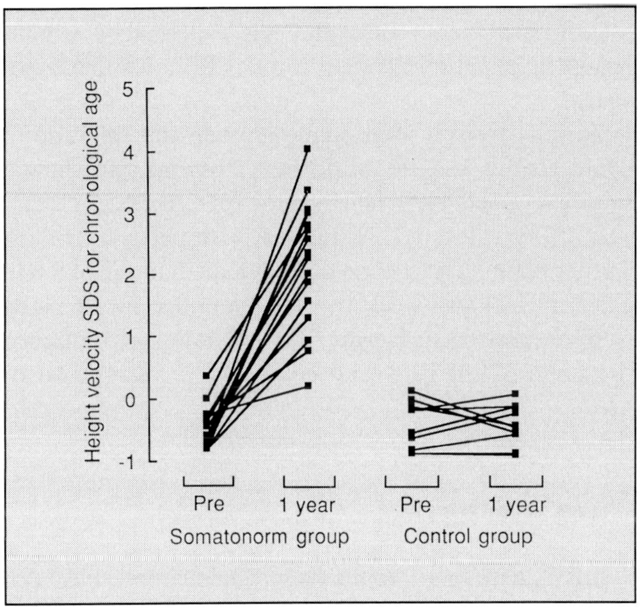

4

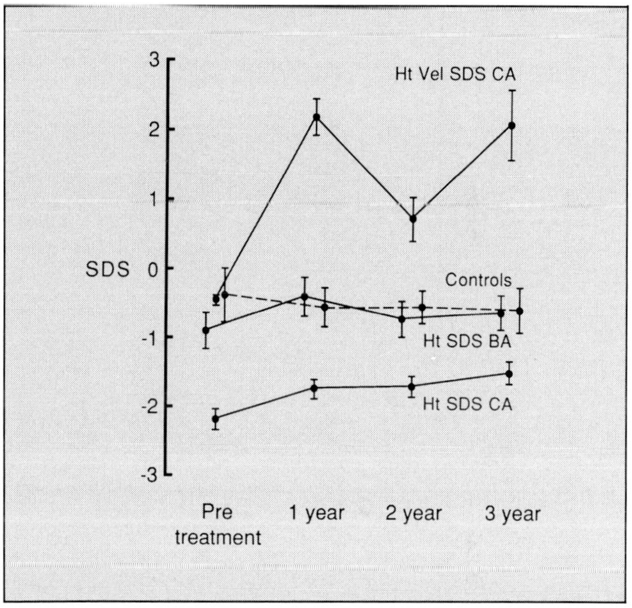

6

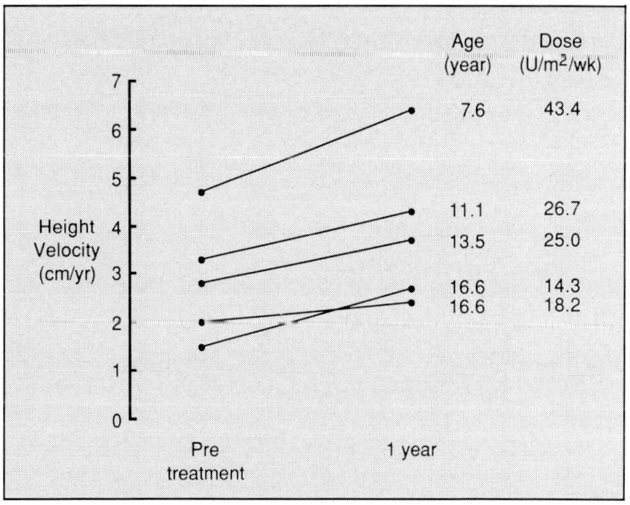

7

Fig. 3. Relationship between gain in height velocity and hGH secretion [18]. AUC$_b$ = Basal area under the curve.

Fig. 4. Changes in height velocity SDS for chronological age in 16 short normal children receiving somatrem compared with 10 controls [15].

Fig. 5. Relation between dose of hGH and change in height velocity SDS over 1 year [15].

Fig. 6. Auxological parameters in 16 short normal children receiving hGH over 3 years (solid lines). Dashed line shows height velocity (Ht Vel) SDS for bone age (BA) in 10 controls. CA = Chronological age.

Fig. 7. Relation between dose of hGH and height velocity increment in Turner girls [1].

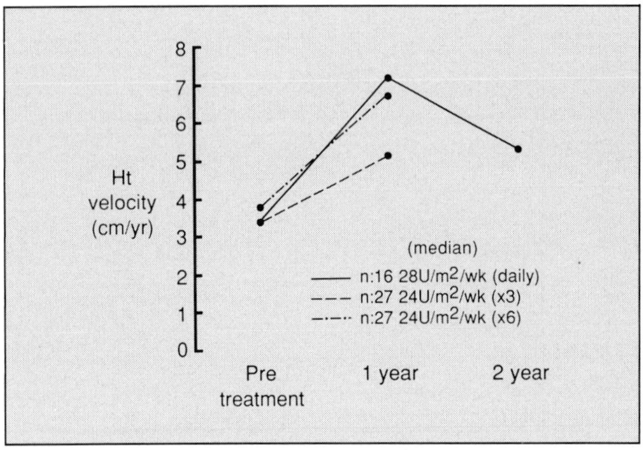

8

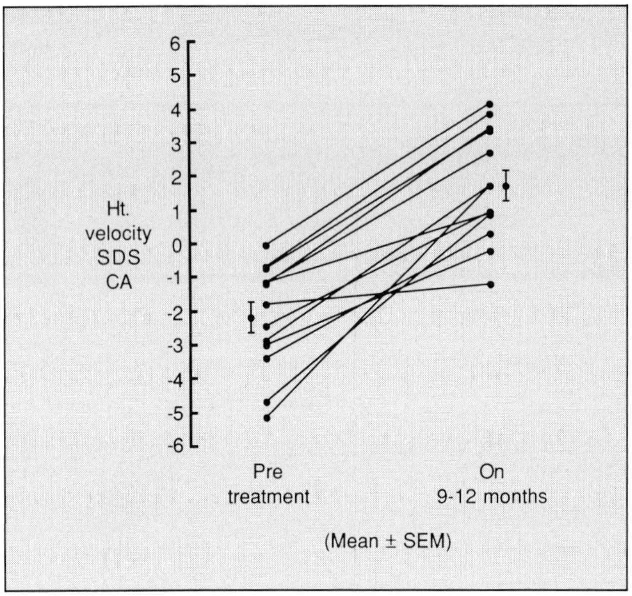

9

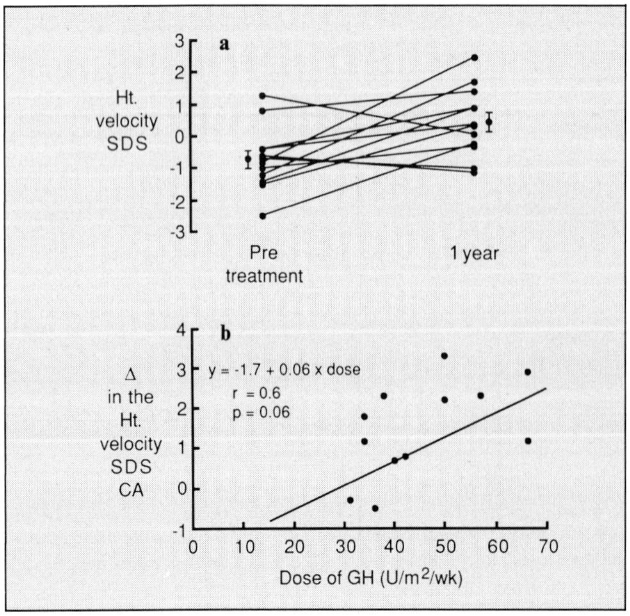

10

Rongen-Westerlaken et al.'s [25] 1st-year results on hGH treatment of 16 girls with Turner syndrome are encouraging and the growth velocity has been maintained in the 2nd year. hGH given in a dose of 28 U/m^2/week on a daily basis has resulted in an increase in height velocity from 3.4 to 7.2 cm/year. There was no unwanted bone age advancement and height prediction improved. 4 of the patients were treated with oestrogen in addition. In another group of 54 Turner girls given 24 U/m^2/week of hGH, the group receiving treatment 6 times weekly (n = 27) grew more quickly than the group treated 3 times weekly (n = 27), once more demonstrating the effect of frequency of administration. The 2nd-year growth rate was higher (5.3 cm/year) than the pre-treatment values (fig. 8). Preliminary results of the Belgian Turner trial have revealed an increase in height velocity from 3.7 to 7.3 cm/year in 20 patients on 1 U/kg/week [26]. Frisch [pers. commun.] has experienced similar results in 20 patients (height velocity from 3.9 to 6.3 cm/year).

Results from a joint study between the Institute of Child Health and the Middlesex Hospital also show a significant increase in height velocity from 4.5 to 7.8 cm/year (SDS values –1.79 and +1.7, respectively) in 13 Turner patients started on hGH at a median age of 6.6 in a dose of 20 U/m^2/week over 9–12 months (fig. 9). None were on oestrogen. Height SDS for bone age remained the same (–1.4) in 9 who completed 1 year on treatment.

Low Birth Weight

LBW, with or without dysmorphic features, has been associated with a lack of catch-up growth and a compromised adult height of –3.6 SDS [27]. In the past, treatment of such children with hGH has not proved to be satisfactory. Although large doses of hGH (up to 66 U/m^2/week) were used in Tanner et al.'s study in 12 chil-

Fig. 8. Change in height (Ht) velocity of Turner girls on hGH treatment over 2 years [25].

Fig. 9. Changes in height (Ht) velocity SDS according to chronological age (CA) in Turner girls on hGH treatment (Middlesex Hospital).

Fig. 10. a Changes in height (Ht) velocity SDS for chronological age (CA) in LBW children on hGH treatment [1]. **b** Relation between growth response and dose of hGH in the same study.

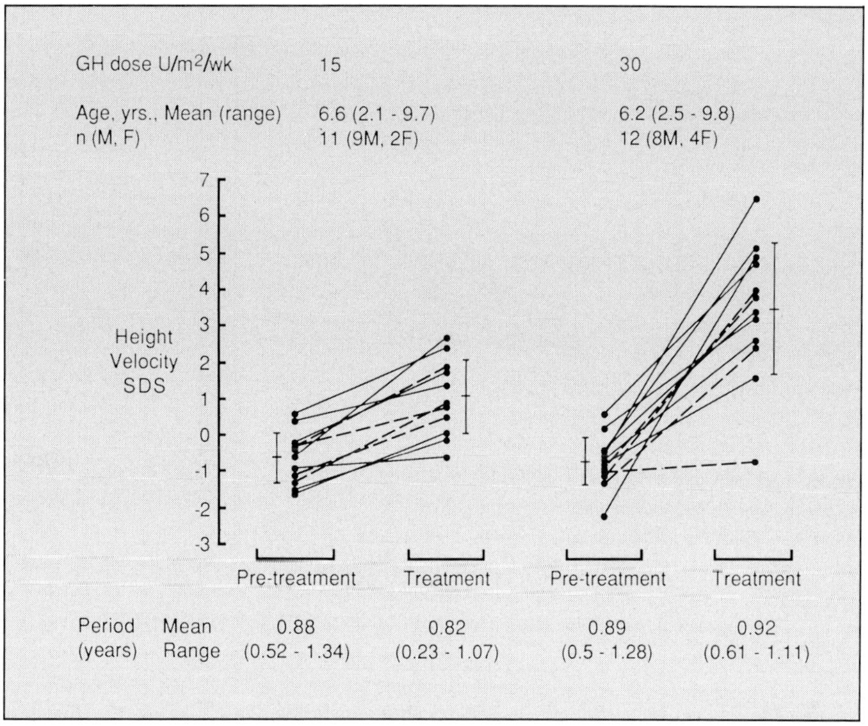

Fig. 11. Changes in height velocity SDS in LBW on hGH treatment: comparison of 2 doses [28]. •——• = LBW with features of Russell-Silver syndrome; •– – –• = no dysmorphic features.

dren with LBW [1], no satisfactory net responses were obtained, with great individual variation. A subtle dose-response correlation is apparent (fig. 10), but a twice weekly injection regimen must certainly have been a drawback.

In a recent study, Stanhope et al. [28] reported a significant increase in height velocity in 20 children with LBW treated with hGH at either 15 or 30 U/m^2/week for 9–12 months. The increment was greater with the larger dose, as shown in figure 11. The bone age advancement, however, was disproportionately greater than the height improvement, although the natural history of skeletal maturation in LBW children is far from clear. There are reports of single cases of successful [Zachmann, pers. commun.] or unsuccessful [29; also Frisch, pers. commun.] treatment. Long-term data are needed, however, to draw conclusions.

Constitutional Delay of Growth and Puberty

It has been recognised for many years that children with constitutional delay of growth and puberty (CDGP) show a transient hGH suppression which has led several clinicians to use hGH in their treatment [30–32]. These children were generally started on treatment with a diagnosis of hGH deficiency based on lack of response to pharmacological tests and their height velocities increased, even on low doses of hGH (8–12 U/week). A decrease in physiological hGH secretion has been shown to occur in CDGP which continues up to pubertal stage 3 [33]. The use of hGH before puberty in these children is thus akin to its use in short normals and the velocity of the childhood component of growth in the late prepubertal years is so slow that heroic doses would be needed to be effective. It is, therefore, to be presumed that hGH would not be helpful in treating children with delayed puberty, even though they may have transient GH deficiency.

To test this hypothesis we have carried out a study comparing the effects of oxandrolone (n = 15) and hGH (n = 11) in children with CDGP aged between 12 and 15 years. Both regimens resulted in a significant height velocity increment (from 3.8 to 8.3 and 6.3 cm/year respectively). Post-treatment values, however, showed a significant increment of mean growth velocity in the oxandrolone group (4.4 cm/year) with respect to the hGH group (3 cm/year) [34]. This sustained effect of oxandrolone makes it superior to hGH. Since hGH is not the major determinant of the pubertal component of growth, it should not be the first choice of treatment for a failure at this stage.

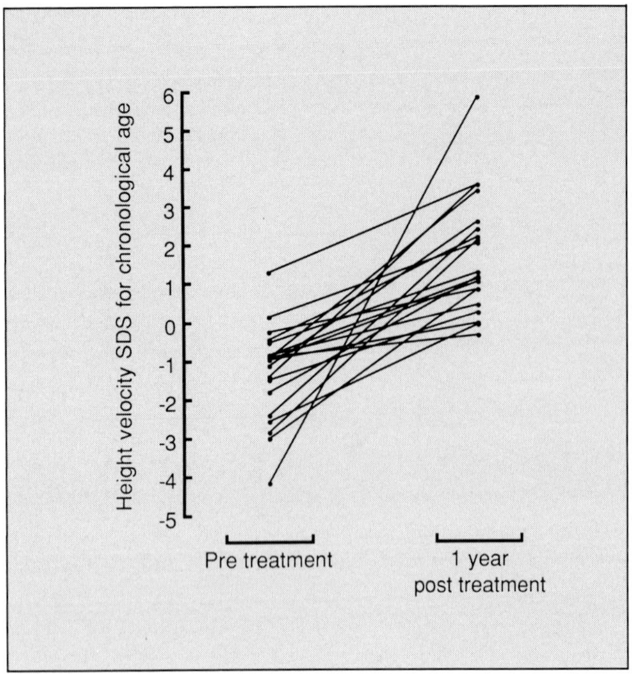

Fig. 12. Changes in height velocity SDS for chronological age in 20 children with hypochondroplasia on hGH treatment [35].

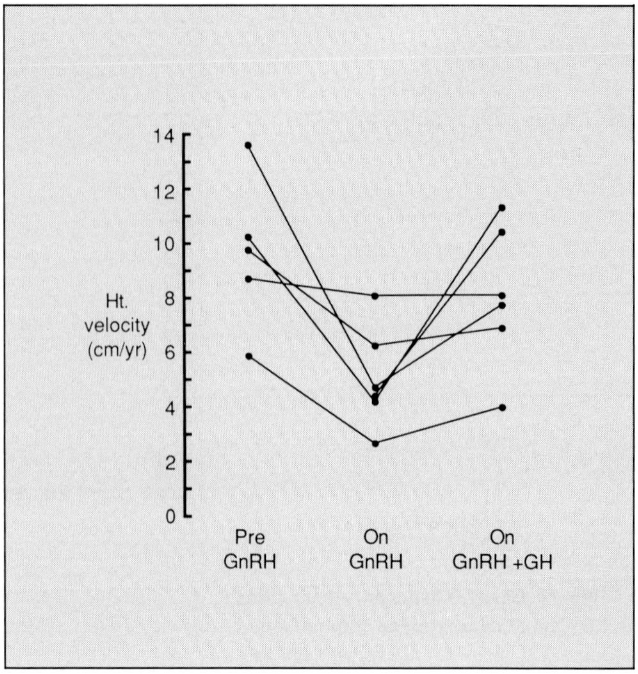

Fig. 13. Changes in height (Ht) velocity in 5 girls with central precocious puberty on GnRH analogue and hGH treatment (Middlesex Hospital).

Hypochondroplasia and Achondroplasia

Hypochondroplasia appears to be a common cause of short stature in our experience. There is a wide spectrum of severity and mild cases are easily missed since the disproportion becomes obvious only in puberty [35]. To date, there has been no medical treatment to improve height in these patients. We treated 20 such children with normal hGH responses to insulin tolerance tests, 15 of whom were prepubertal, with hGH in doses between 12 and 32 U/m^2/week on a nightly basis for 1 year. A significant acceleration in growth rate occurred, mean height velocity increasing from –1.66 to +1.62 (fig. 12). Sitting height SDS and subischial leg length SDS also showed significant increases proportionately with no unwanted bone age advancement.

2 children with achondroplasia and 3 with spondyloepiphyseal dysplasia are currently being treated with hGH in doses between 20 and 30 U/m^2/week for 9–12 months at the Middlesex Hospital. The preliminary results are encouraging, with height velocity SDS increasing from –1.7 to –0.3 in the former group and from –1.6 to +0.14 in the latter. A longer period will be required to assess the effects of treatment on adult height prognosis.

Precocious Puberty

Gonadotrophin-releasing hormone (GnRH) analogues are now being successfully used in the treatment of gonadotrophin-dependent precocious puberty. The suppression of gonadotrophin pulsatility and sexual maturation are accompanied by a decrease in growth rate. Height prognoses of such patients on treatment, however, do not show a significant alteration [36]. It has been reported that physiological hGH secretion associated with decreased sex steroid secretion on GnRH treatment is also decreased [37], which may account for the poor growth and lack of improvement of height prognosis. A possible therapeutic manoeuvre would be to treat such children with hGH as well as a GnRH analogue.

At the Middlesex Hospital we have studied the effect of hGH on the growth of 9 girls with central precocious puberty treated with GnRH analogue for 3 years. Pubertal development was arrested but growth rate had fallen from 9.6 to 5.0 cm/year. hGH given daily at a dose of 20 U/m^2/week over 0.2–1.1 years increased growth velocity from 5.0 to 8.4 cm/year ($p = 0.03$; fig. 13). The increase in height was equal to or greater than the increase in bone age in the patients who have completed a year on treatment. This implies a better predicted adult height.

Wit started a similar trial in November 1988 [pers. commun.], the results of which are preliminary at the moment.

Other Chronic Diseases

The uses of hGH in chronic diseases are being evaluated in current trials. 6 patients with chronic renal failure and 12 patients (6 prepubertal) with renal transplant and growth retardation are being treated with hGH at Guy's Hospital [Lesley Rees, pers. commun.]. The preliminary results following 6 months of treatment have yielded a significant increase in height velocity, the greatest response being in the renal failure group. The long-term benefits of therapy in this group of children are under investigation.

Wit's results on renal failure will be available shortly. Vanderschueren-Lodeweyckx's GH trial in 20 children with renal failure has completed its first 6 months [pers. commun.]. One single case has been treated successfully by Frisch with a doubling of height velocity [pers. commun.].

Analysis of growth data of asthmatic children has demonstrated a normal height velocity up to 10 years of age followed by a delayed onset of puberty and growth deceleration. The ultimate height was found to be within the parental height range [38]. We have treated 1 boy aged 9.3 years with hGH (16 U/m^2/week) for 1 year. Height velocity SDS increased from -2.54 to $+2.15$. However, a detailed study of endocrine investigations and potential use of hGH in asthmatic children has yet to be undertaken.

A few cases of familial short stature have not responded to treatment [Laron, Zachmann, pers. commun.].

3 cases of Fanconi's anaemia with poor growth rate and no hGH response to pharmacological tests were treated successfully for 1 year by Aynsley-Green et al. [39].

Another interesting use of hGH may be in the field of reproductive endocrinology. In a study carried out by Jacobs' group at the Middlesex Hospital [40], the addition of hGH to treatment with human menopausal gonadotrophin (hMG) significantly augmented the ovarian response in 4 patients treated for fertilisation who have been previously resistant to hMG. The basis of this study lies in the recent observations that hGH enhances granulosa cell production of insulin-like growth factor I (IGF-I) and IGF-I in turn has been found to act synergistically with follicle-stimulating hormone to modulate ovarian function.

Conclusions

It is evident from these studies that hGH is far more successful in promoting growth in a variety of conditions than earlier reports suggested. The better response is probably due to more frequent administration of hGH in doses appropriate for body size.

All children respond to hGH treatment with an increased growth rate, the magnitude of which depends on the pre-treatment height velocity and dose of hGH administered. However, patients with Turner syndrome, skeletal dysplasias or other conditions may need more hGH than normal children or children with hypopituitarism.

Although results on the ultimate height of patients receiving hGH in previous studies were not very promising, hGH given in sufficient dosage for long enough must increase final height. If it did not do so tall adults in general and pituitary giants in particular would not exist.

Acknowledgements

We thank the following colleagues for access to their unpublished data: K. Albertsson-Wikland, C. Chantler, H. Frisch, Z. Laron, J. Perheentupa, M.A. Preece, M. Ranke, L. Rees, R. Stanhope, M. Vanderschueren-Lodeweyckx, J. Wit, and M. Zachmann.

References

1 Tanner JM, Whitehouse RH, Hughes PCR, et al: Effect of human growth hormone treatment for 1 to 7 years on growth of 100 children, with growth hormone deficiency, low birth weight, inherited smallness, Turner's syndrome and other complaints. Arch Dis Child 1971;46:745–780.

2 Hindmarsh PC, Matthews DR, Brook CGD: Growth hormone secretion in children determined by time series analysis. Clin Endocrinol 1988;29:35–44.

3 Albertsson-Wikland K, Rosberg S: Analysis of 24-hour growth hormone profiles in children: Relation to growth. J Clin Endocrinol Metab 1988;67:1–8.

4 Mauras N, Blizzard RM, Link K, et al: Augmentation of growth hormone secretion during puberty: Evidence for a pulse amplitude-modulated phenomenon. J Clin Endocrinol Metab 1987; 64:596–601.

5 Stanhope R, Pringle PJ, Brook CGD: The mechanism of the adolescent growth spurt induced by low dose pulsatile GnRH treatment. Clin Endocrinol 1988;28:83–91.
6 Kastrup KW, Christiansen JS, Anderson JK, et al: Increased growth rate following transfer to daily sc administration from three weekly im injections of hGH in growth hormone deficient children. Acta Endocrinol 1983;104:148–152.
7 Hindmarsh P, Smith PJ, Brook CGD, et al: The relationship between height velocity and growth hormone secretion in short prepubertal children. Clin Endocrinol 1987;27:581–591.
8 Spadoni GL, Cianfarani S, Bernardini S, et al: Twelve hour spontaneous nocturnal growth hormone secretion in growth retarded patients. Clin Pediatr 1988;10:473–478.
9 Spiliotis BE, August GP, Hung W, et al: Growth hormone neurosecretory dysfunction. JAMA 1984;251:2223–2330.
10 Hindmarsh PC, Brook CGD: Auxological and biochemical assessment of short stature. Acta Paediatr Scand 1988;343(suppl): 73–76.
11 Rose SR, Ross JL, Uriarte M, et al: The advantage of measuring stimulated as compared with spontaneous growth hormone levels in the diagnosis of growth hormone deficiency. N Engl J Med 1988;319:201–207.
12 Smith J, Hindmarsh PC, Brook CGD: Contribution of dose and frequency of administration to the therapeutic effect of growth hormone. Arch Dis Child 1988;63:491–494.
13 Wit JM, Faber JAJ, Van Den Brande JL: Growth response to human growth hormone treatment in children with partial or total growth hormone deficiency. Acta Paediatr Scand 1986;75: 767–773.
14 Hindmarsh PC, Smith PJ, Pringle PJ, et al: The relationship between the response to growth hormone therapy and pretreatment growth hormone secretory status. Clin Endocrinol 1988; 28:559–563.
15 Hindmarsh PC, Brook CGD: Effect of growth hormone on short normal children. Br Med J 1987;295:573–577.
16 Darendeliler F, Hindmarsh PC, Brook CGD: Dose-response curves for growth hormone. J Endocrinol 1990;125:311–316.
17 Buchanan CR, Law CM, Milner RDG: Growth hormone in short slowly growing children and those with Turner's syndrome. Arch Dis Child 1987;62:912–916.
18 Albertsson-Wikland K: Growth hormone treatment in short children: Short-term and long-term effects on growth. Acta Paediatr Scand 1988;343(suppl):77–84.
19 Wit JM, Rietveld HF, Drop SLS, et al: A controlled trial of methionyl growth hormone therapy in prepubertal children with short stature, subnormal growth rate and normal growth hormone response to secretagogues. Acta Paediatr Scand 1989;78: 426–435.
20 Bozzola M, Cisternino M, Biscaldi I, et al: Effectiveness of growth hormone (GH) therapy in GH-deficient children and non-GH-deficient short children. Eur J Pediatr 1988;147:248–251.
21 Stahnke N: hGH treatment in short children without growth hormone deficiency. N Engl J Med 1984;310:925–926.
22 Brook CGD: Growth hormone deficiency in Turner's syndrome. N Engl J Med 1978;298:1203–1204.
23 Bottazzo GP, McIntosh C, Stanford W, et al: Growth hormone cell antibodies and partial GHD in a girl with Turner's syndrome. J Clin Endocrinol 1980;12:1–9.
24 Butenandt O: Growth hormone deficiency and growth hormone therapy in Ullrich-Turner syndrome. Klin Wochenschr 1980;58: 99–101.
25 Rongen-Westerlaken C, Wit JM, Drop SLS, et al: Methionyl hGH in Turner's syndrome. Arch Dis Child 1988;63:1211–1217.
26 Vanderschueren-Lodeweyckx M, Massa G, Maes M, et al: Growth promoting effect of GH and low dose ethinyl oestradiol in girls with Turner's syndrome. J Clin Endocrinol Metab 1990; 70:122–126.
27 Davies PSW, Valley R, Preece MA: Adolescent growth and pubertal progression in the Silver-Russell syndrome. Arch Dis Child 1988;63:130–135.
28 Stanhope R, Ackland F, Hamill G, et al: Physiological growth hormone secretion and response to GH treatment in children with short stature and IUGR. Acta Paediatr Scand 1989; 349(suppl):47–52.
29 Lenko HL, Leisti S, Perheentupa J: The efficacy of GH in different types of growth failure. Eur J Pediatr 1982;138:241–249.
30 Prader A, Zachmann M, Bucher H: Constitutional delay of growth and puberty: Auxological and endocrine characteristics; in Cacciari E, Prader A (eds): Pathophysiology of Puberty. Serono Symposia. London, Academic Press, 1980, vol 36, pp 123–135.
31 Bierich JR, Potthoff K: Die Spontansekretion des Wachstumshormons bei der konstitutionellen Entwicklungsverzögerung und der frühnormalen Pubertät. Monatsschr Kinderheilk 1979; 127:561–565.
32 Trygstad O: Transitory growth hormone deficiency successfully treated with human growth hormone. Acta Endocrinol 1977;84: 11–22.
33 Bierich JR: Treatment by hGH of constitutional delay of growth and adolescence. Acta Paediatr Scand 1986;325(suppl):71–76.
34 Büyükgebiz A, Hindmarsh P, Brook CGD: Oxandrolone versus GH treatment of CDGP. Arch Dis Child 1990;65:448–452.
35 Appan S, Laurent S, Chapman M, et al: Growth and growth hormone therapy in hypochondroplasia. Acta Paediatr Scand, in press.
36 Stanhope R, Pringle PJ, Brook CGD: Growth hormone and sex steroid secretion in girls with central precocious puberty treated with a GnRH analogue. Acta Paediatr Scand 1988;77:525–530.
37 Mansfield MJ, Rudlin CR, Crigler JF, et al: Changes in growth and serum growth hormone and plasma SmC levels during suppression of gonadal sex steroid secretion in girls with central precocious puberty. J Clin Endocrinol Metab 1988;66:3–9.
38 Balfour-Lynn L: Growth and childhood asthma. Arch Dis Child 1986;61:1049–1055.
39 Aynsley-Green A, Zachmann M, Werder EA, et al: Endocrine studies in Fanconi's anaemia. Arch Dis Child 1978;53:126–131.
40 Homburg R, Eshel A, Abdulla HI, et al: Growth hormone facilitates ovulation induction by gondotrophins. Clin Endocrinol 1988;29:113–117.

Prof. C.G.D. Brook
Endocrine Unit
The Middlesex Hospital
Mortimer Street
London W1N 8AA (UK)

Horm Res 1990;33:137–143

0301–0163/90/0334–0137$2.75/0

Non-Conventional Growth Hormone Therapy in Turner Syndrome: The United States Experience

R.G. Rosenfeld

Department of Pediatrics, Stanford University School of Medicine, Stanford, Calif., USA

Key Words. Growth hormone · Turner syndrome · Growth · Short stature

Abstract. Beginning in 1983, 71 girls with Turner syndrome were enrolled in a prospective, randomized study of human growth hormone (hGH), alone and in combination with oxandrolone; 4–5 patient-year data are currently available on 66 subjects. While the annual growth rate of the untreated control group (expressed as a Z score for untreated Turner girls) was –0.1 SD, the annual Turner growth rate Z scores for girls receiving hGH alone were +3.1, +2.0, +1.5 and +2.9 SD for years 1–4, respectively (hGH was given daily in year 4). Mean annual Turner growth rate Z scores for subjects on combination therapy with oxandrolone were +6.6, +4.3, +3.0 and +2.7 SD for years 1–4, respectively. The mean height for the 20 subjects ≥ 16 years of age is 150 cm.

Although the pathogenesis of growth failure in Turner syndrome remains unknown, short stature is found in over 95% of diagnosed patients [1–6]. A recent compilation of data from 4 major European studies has reported a final adult height of 143.1 cm [7], consistent with data from most American studies.

The development of recombinant DNA-derived human growth hormone (hGH) has made possible a series of prospective studies of hGH use in patients with Turner syndome [8–10]. Such investigations are currently in progress in Europe, Australia, Canada, Japan and South America, as well as in the United States, and preliminary results have been remarkably concordant [reviewed in ref. 11]. The earliest of these studies, begun in the United States in 1983, is now entering its 7th year, with data already available on subjects who have been treated for 4–5 years [8–10].

Subjects and Methods

Seventy-one girls with Turner syndrome, ranging in age from 4.4 to 12.4 years, were enrolled. Three subjects were withdrawn from the study during the first 12 months. Subjects were divided among the following groups: group 1 (n = 18): no treatment for a period of 12–21 months, followed by allocation to group 5; group 2 (n = 19): oxandrolone, 0.125 mg/kg/day for 12–21 months, followed by allocation to group 5; group 3 (n = 17): methionyl hGH, 0.125 mg/kg intramuscularly 3 times/week; after 3 years of treatment, subjects were placed on daily therapy, but with the same cumulative weekly dosage of hGH; group 4 (n = 17): combination oxandrolone, 0.125 mg/kg/day, and met-hGH, 0.125 mg/kg intramuscularly 3 times/week. After 12–21 months of treatment, subjects were allocated to group 5; group 5 (n = 49): this group is comprised of subjects formerly in groups 1, 2 and 4. After completion of their first phase of therapy, all these subjects were placed on combination oxandrolone, 0.0625 mg/kg/day, and met-hGH, 0.125 mg/kg intramuscularly 3 times/week (phase 2). After 24 months in phase 2 of treatment, subjects were randomized to combination therapy with met-hGH given 3 times/week or daily, employing the same cumulative weekly dosage. (Editors comment: the mean equivalent of 1 mg is 2 units, although there are slight differences between preparations.)

Results

Table 1 presents the growth rates for each group. Growth rates for all treatment groups were significantly greater for the first 2 years, when compared with either pretreatment growth rates or the growth rate of group 1 (control) during year 1. However, since the normal growth pattern of Turner patients is one of progressive growth deceleration, it is more informative to compare

Table 1. Growth rates for Turner subjects

Group	n	Treatment	Dosage	Growth rate, cm/year				
				pretreatment	year 1	year 2	year 3	year 4
1	18	no treatment		4.2 ± 1.1	3.8 ± 1.1			
1 → 5		oxandrolone + hGH				8.3 ± 1.2	6.7 ± 1.4	
2	19	oxandrolone	0.125 mg/kg/day	4.1 ± 1.9	7.6 ± 1.5			
2 → 5		oxandrolone + hGH				7.1 ± 1.6	5.3 ± 2.4	
3	17	hGH	0.125 mg/kg/3 t.i.w.	4.5 ± 0.8	6.6 ± 1.2	5.4 ± 1.1	4.6 ± 1.4	
3		hGH	daily					5.5 ± 1.5
4	17	oxandrolone + hGH	0.125 mg/kg/day 0.125 mg/kg/3 t.i.w.	4.3 ± 0.9	9.8 ± 1.4			
4 → 5	49	oxandrolone + hGH	0.0625 mg/kg/day 0.125 mg/kg/3 t.i.w.			7.4 ± 1.4	6.1 ± 1.5	4.9 ± 1.5
4 → 5		oxandrolone + hGH	0.0625 mg/kd/day daily					5.3 ± 1.7

Groups 1, 2, 4: after 12–21 months, allocated to group 5. Results are mean ± SD. t.i.w. = times/week.

Table 2. Height velocity standard deviation score for Turner syndrome

Group	n	Treatment	Dosage	Height velocity standard deviation score				
				pretreatment	year 1	year 2	year 3	year 4
1	18	no treatment		0.2 ± 1.2	−0.1 ± 1.0			
1 → 5		oxandrolone + hGH				5.5 ± 1.4	4.0 ± 1.7	
2	19	oxandrolone	0.125 mg/kg/day	0.2 ± 1.0	4.4 ± 1.8			
2 → 5		oxandrolone + hGH				4.2 ± 1.5	2.4 ± 2.3	
3	17	hGH	0.125 mg/kg/3 t.i.w.	0.5 ± 0.8	3.1 ± 1.2	2.0 ± 1.1	1.5 ± 1.4	2.9 ± 1.4
4	17	oxandrolone + hGH	0.125 mg/kg/day 0.125 mg/kg/3 t.i.w.	0.2 ± 0.9	6.6 ± 1.2			
4 → 5	49	oxandrolone + hGH	0.0625 mg/kg/day 0.125 mg/kg/3 t.i.w.			4.3 ± 1.4	3.0 ± 1.4	2.7 ± 1.3

Results are mean ± SD for height velocity for untreated Turner patients, based upon the data of Ranke et al. [6]. Groups 1, 2, 4: after 12–21 months, allocated to group 5. t.i.w. = times/week.

subject growth rates with those of untreated Turner girls, employing the data of Ranke et al. [6] (table 2).

At the time of this writing (autumn, 1989), of the 16 girls in group 3 (met-hGH alone), 15 (94%) have heights above the mean for untreated Turner girls, 7 (44%) have heights above the 90th percentile for untreated Turner girls, and 9 (56%) have already exceeded their projected adult heights. Of the 49 girls originally in groups 1, 2 and 4, and subsequently on combination therapy in group 5, 47 (96%) have heights above the mean for untreated Turner girls, and 35 (72%) have heights above the 90th percentile for untreated Turner girls. When all groups are combined, 20 girls are now at chronological ages ≥ 16 years. Although many are still growing, their mean height has already reached 150 cm.

Discussion

The first paper describing treatment of Turner syndrome with hGH appeared in 1960 [12]. Over the ensuing 25 years, a number of reports appeared in the literature. These studies were characterized by small numbers of subjects, brief duration of treatment, low dosages of hGH and lack of randomization [for review, see ref. 11, 13]. It was only with the recent development of methodologies allowing for the 'large-scale' manufacture of recombinant hGH that it became possible to perform adequate randomized, prospective trials.

Although the American study reported here is still in progress, several conclusions can be justifiably made at this time.

(1) hGH alone and, in combination with oxandrolone, can stimulate skeletal growth for at least 3 years. When daily administration of hGH is employed, significant stimulation of growth can be seen for at least 4 years. These conclusions are best supported when data are compared to the typical growth patterns exhibited by untreated Turner patients.

(2) In addition to stimulation of short-term growth, an effect of therapy upon final height seems probable. The projected heights of our subjects at time of entry into the study, based upon the growth curves of Lyon et al. [7], indicate a mean final height of 143 cm, similar to the mean adult height of 143 cm predicted by the method of Bayley and Pinneau [14]. However, the majority of subjects have already exceeded their projected adult heights, even though most are still growing and receiving treatment.

(3) The growth response is, however, a modest one. Turner patients do not respond to hGH in the same manner as do patients with classical hGH deficiency. Nevertheless, it seems reasonable to expect, at this point, that early diagnosis of Turner syndrome, combined with initiation of hGH and related therapy, may allow many girls to reach final heights $\geq$ 150 cm, and also permit the initiation of estrogen replacement at a more appropriate age.

Several recent studies have provided supporting data for the use of hGH in Turner syndrome, and have provided alternative regimens to those employed in the American study. Rongen-Westerlaken et al. [15] employed daily administration of hGH at a dosage of 0.9 IU/kg/week, while Vanderscheuren-Lodeweyckx et al. [16] used daily hGH at a dosage of 1.0 IU/kg/week. Annual growth rates increased from 3.4 and 3.7 cm, respectively, to 7.2 and 7.3 cm, respectively. In an extension of the American study, Rosenfeld et al. [unpubl. results] employed daily hGH at a dosage of 0.375 mg/kg/week (the same cumualtive weekly dosage as employed in the study reported in this paper), and observed a 1st-year growth rate of 7.9 cm.

Nilsson et al. [17] employed combination hGH and oxandrolone therapy in Swedish Turner patients, using dosages similar to those employed by Rosenfeld et al. [8, 9], but with daily administration of hGH. The annual growth rate increased from 3.9 to 9.4 and 6.8 cm/year in years 1 and 2, respectively. These growth rates are similar to those reported by Rosenfeld et al. [8, 9], although the Turner height velocity Z scores are somewhat higher (+7.6 and +5.1 in years 1 and 2, respectively), due to the slightly older age of their subjects (mean pretreatment age 12.3 years).

Although initial studies combining estrogen and hGH therapy have not been particularly promising, it is conceivable that the optimal dosages of estrogen and age of administration have not yet been determined [18]. Alternatively, it may be that estrogen therapy has no place in treatment designed to stimulate growth in Turner patients, but still may be introduced at an earlier age than is currently customary for feminization. Initiation of growth-promoting therapy at an early age may normalize childhood growth sufficiently to permit replacement estrogen therapy by 12–13 years. Clinical trials combining hGH and low-dose estrogen in adolescent Turner patients (age 12–15 years) are currently beginning in the United States.

References

1 Palmer CG, Reichmann A: Chromosomal and clinical findings in 110 females with Turner syndrome. Hum Genet 1976;35:35–49.

2 Park E, Bailey JD, Cowell CA: Growth and maturation of patients with Turner's syndrome. Pediatr Res 1983;17:1–7.

3 Brook CGD, Murset G, Zachmann M, Prader A: Growth in children with 45,XO Turner's syndrome. Arch Dis Child 1974;49: 789–795.

4 Lev-Ran A: Androgens, estrogens, and the ultimate height in XO gonadal dysgensis. Am J Dis Child 1977;131:648–649.

5 Sybert VP: Adult height in Turner syndrome with and without androgen therapy. J Pediatr 1984;104:365–369.

6 Ranke MB, Pfluger H, Rosendahl W, et al: Turner syndrome: Spontaneous growth in 150 cases and review of the literature. Eur J Pediatr 1983;141:81–88.

7 Lyon AJ, Preece MA, Grant DB: Growth curve for girls with Turner syndrome. Arch Dis Child 1985;60:392–395.

8 Rosenfeld RG, Hintz RL, Johanson AJ, et al: Methionyl human growth hormone and oxandrolone in Turner syndrome: Preliminary results of a prospective randomized trial. J Pediatr 1986; 109:936–943.

9 Rosenfeld RG, Hintz RL, Johanson AJ, et al: Results from the first 2 years of a clinical trial with recombinant DNA-derived human growth hormone (somatrem) in Turner's syndrome. Acta Paediatr Scand [Suppl] 1987;331:59–66.

10 Rosenfeld RG, Hintz RL, Johanson AJ, et al: Three-year results of a randomized prospective trial of methionyl human growth hormone and oxandrolone in Turner syndrome. J Pediatr 1988; 113:393–400.

11 Rosenfeld RG: Update on growth hormone therapy for Turner's syndrome. Acta Paediatr Scand [Suppl] 1989;356:103–108.

12 Escamilla RF, Hutchings JJ, Deamer WC, et al: Clinical experiences with human growth hormone (LI) in pituitary infantilism and in gonadal dysgenesis (abstract). 1st Int Congr Endocrinol, Copenhagen, July 1960. Acta Endocrinol 1960;35(suppl 51):253.
13 Wilton P: Growth hormone treatment in girls with Turner's syndrome: A review of the literature. Acta Paediatr Scand 1987;76: 193–200.
14 Bayley N, Pinneau SR: Tables for predicting adult height from skeletal age; Revised for use with the Greulich-Pyle hand standards. J Pediatr 1952;40:423–441.
15 Rongen-Westerlaken C, Wit JM, Drop SLS, et al: Methionyl human growth hormone in Turner's syndrome. Arch Dis Child 1988;63:1211–1217.
16 Vanderscheuren-Lodeweyckx M, Massa G, Maes M, et al: The Belgian Genotropin Turner Trial: Preliminary results (abstract). 6th Int Symp Growth Disorders, Brussels, Oct 1988. Acta Paediatr Scand [Suppl] 1989;349:158.
17 Nilsson KO, and the Swedish Paediatric Study Group for Growth Hormone Treatment: The Swedish Somatonorm Turner Trial: Two-year results (abstract). Acta Paediatr Scand [Suppl], 1989;356:160.
18 Ross JL, Cassorla F, Carpenter G, et al: The effect of short term treatment with growth hormone and ethinyl estradiol on lower leg growth rate in grils with Turner's syndrome. J Clin Endocrinol Metab 1988;67:515–518.

Dr. R.G. Rosenfeld
Department of Pediatrics
Stanford University School of Medicine
Stanford, CA 94305 (USA)

Discussion

J.C. Job, France: In addition to the data presented in the Poster session and to those collected and commented on by Dr. Charles Brook, I wish to summarise the 1-year results of a double-blind placebo-controlled multicentre clinical trial with hGH (Sanofi-Choay, France) in 70 pre-pubertal children (42 M, 28 F) having a very short stature due to intra-uterine growth retardation (IUGR). Their mean chronological age at the onset of the trial was 8.8 ± 2.1 (SD) years, bone age was 6.1 ± 2.1 years and mean height SDS was −3.0 ± 0.8 (SD). They were the so-called true, proportionate idiopathic IUGR defined by a birth length below −2 SD for gestational ages 36–41 weeks (−3.4 ± 1.1 SDS; mean ± SD), a birth weight also below −2 SD for gestational age, no obvious malformation or other known cause for IUGR, and a hGH response above 10 ng/ml during at least 2 recent stimulation tests. They were followed in 11 paediatric clinics in France and Belgium. They were assigned by randomisation to 2 different doses of hGH, 0.4 IU/kg/week (group A, n = 35) or 1.2 IU/kg/week (group B, n = 35), divided in 6 daily subcutaneous injections administered between 4 and 8 p.m. 40% of the children received daily subcutaneous injections of placebo during the first 6 months, then were switched to hGH, either 0.4 (group A, n = 14) or 1.2 IU/kg/week (group B, n = 15), for the following 6 months. Ethical approval and informed consent from children and/or parents were obtained. The whole cohort has been followed for 12 months, without detectable side-effects nor major psychological difficulties regarding both the injections themselves and the acceptance of a blind randomised study for the 1st year of treatment. The results (mean ± SD) have been as follows:

		Annual growth rate		Height at the end of the period, SDS for age
		SDS for age	cm/year	
Pre-treatment year	Group A	−0.57 ± 1.12	5.16 ± 1.20	−2.90 ± 0.75
	Group B	−0.58 ± 1.00	5.08 ± 1.03	−3.02 ± 0.72
	Group placebo	−1.06 ± 1.46	4.55 ± 1.45	−3.23 ± 0.87
0–6 months	Group A	+0.99 ± 1.55	6.76 ± 1.63	−2.71 ± 0.85
	Group B	+4.25 ± 3.10	9.13 ± 2.51	−2.60 ± 0.70
	Group placebo	−0.58 ± 2.23	5.04 ± 2.08	−3.19 ± 0.87
7–12 months	Group A	+1.32 ± 1.70	7.03 ± 1.56	−2.49 ± 0.80
	Group B	+3.48 ± 2.75	8.55 ± 2.07	−2.26 ± 0.71
	From placebo to A	+1.38 ± 1.84	6.86 ± 1.80	−2.76 ± 1.04
	From placebo to B	+4.37 ± 2.90	9.29 ± 2.30	−2.99 ± 0.85

All changes within hGH-treated groups were highly significant. The 6-month changes with placebo were not.

Bone age was evaluated independently from the investigators by 2 skilled and independent paediatric radiologists, using the method of Greulich and Pyle. At present we have the 6-month bone age data for all 70 patients, and the 12-month data for only two thirds of them. During this time, no one has accelerated bone maturation. After 6 months the changes in height age to bone age ratio have been 0.00 ± 0.03 (SD) in group A, 0.03 ± 0.05 (SD) in group B and −0.02 ± 0.08 in the children having received placebo, none being significant.

These preliminary data seem encouraging. The trial is to be continued for a 2nd year. The full reults will be presented to the ESPE as soon as they will be available. The Coordinators of the study are Profs. P. Chatelain, J.C. Job and M. Vanderschueren-Lodeweyckx. The other senior investigators are: Profs. Chaussain, Girard, Ponte, Rochiccioli, Toublanc, and Dr. Colle, in France, Profs. Craen, Du Caju and Maes in Belgium. The statistical calculations were done by Drs A. Galland and L. Sagnard (Sanofi Recherche, France).

M.A. Preece, UK: Dr. Brook asserts that changes in height SDS for bone age are more reliable than changes in formal height prediction, by methods such as the TW2 (Tanner-Whitehouse atlas), in assessment of height potential. Published comparisons of these methods have shown the reverse, albeit in different clinical situations. Dr. Brook cannot make this statement until mature height is known when a formal comparison can be made.

C.G.D. Brook, UK: I agree with the last statement but I do not believe in predicting height of abnormal subjects.

W.M. Teller, FRG: Have you looked into the glucose tolerances of your Turner patients? Since Turner patients have an intrinsic tendency to abnormal glucose tolerance, did they deteriorate on long-term hGH treatment?

R. Rosenfeld, USA: Before therapy, Turner subjects had a significantly greater glucose response during OGTT than published normal control values, although none were frankly diabetic. 12 months of hGH therapy did not result in a change in either fasting glucose or integrated glucose concentrations. Oxandrolone alone or in combination with hGH did, however, result in significant increases in the mean integrated insulin concentration. The long-term ramifications of such hyperinsulinemia remain to be established.

J.M. Gertner, USA: Do you think it reasonable to include early pubertal children with constitutional delay in trials of hGH therapy?

C.G.D. Brook: We have tested this hypothesis and hGH is less effective than sex steroids.

Z. Laron, Israel: From the results obtained to date, at what age do you think hGH therapy should be initiated in Turner patients?

R. Rosenfeld: There do not yet exist adequate data to support a recommendation for optimal age for initiation of therapy. However, we should keep several things in mind. (1) Most identified patients with Turner syndrome fall off the normal growth curve by the age of 5 years. Once they have fallen off, it is highly unusual for any of them to return to the normal growth curve on their own. (2) The response of Turner patients to hGH is a modest one, when compared with the response of hGH-deficient patients. However, if therapy is begun before the patient is too far below the normal growth curve, a modest response may be adequate to keep the patient within the broad normal range. Our current policy is to begin therapy as soon as the patient falls below the 5th percentile for chronological age, although we have not begun prior to 2 years of age. In young patients, we have employed hGH alone, given daily. In older patients, where the window of treatment time may be smaller, we have employed combination hGH and oxandrolone.

C.G.D. Brook: You would have to admit, Ron, that logic dictates that the answer to Prof. Laron's question is that, if hGH works at all, it ought to be started before any deficit is incurred just as it should be in hypopituitary children. Dr. Kaplan frequently says that if you treat a hypopituitary child from the day of birth it is bound to end up a normal child and the same must necessarily be true if you believe in the use of hGH in Turner syndrome; to wait until the child is on the 5th height centile for normal girls, in fact you incur a deficit which you probably cannot or will never recover.

P. Boepple, USA: Many of the presentations in these proceedings have appropriately included assessment of the hGH dose and regimen when evaluating the impact of hGH therapy in a variety of clinical settings. It seems similarly inappropriate to make blanket statements regarding the impact of GnRH agonist therapy, since the various agents employed in clinical studies of precocious puberty vary in their potencies by almost a log order and the routes of administration result in 20-fold differences in bioavailability. One should more appropriately consider the impact of GnRH agonist therapy when the goal of pituitary-gonadal suppression has been consistently attained. With that in mind, we have reported significant improvement in the final height prognosis in children with precocious puberty, especially if growth data are analyzed in a developmental context using bone age as an index.

C.G.D. Brook: There is some merit in what you say but, nevertheless, we have compared different analogues and your suggestion is not actually borne out by experience. Our society had a prolonged discussion about this and, in fact, it may not be in the best interests of our patients to castrate them medically but in so suppressing their sex hormones we may prolong the duration of their growth potential.

N. Stahnke, FRG: Treating patients with achondroplasia with hGH, do you not fear to increase the body disproportions in this disease? One short remark on Fanconi's anaemia. In this condition there is a 15% risk of developing leukaemia without any hGH treatment. Does it not seem dangerous to give hGH to these patients?

C.G.D. Brook: Thus far, the body proportions of patients with achondroplasia treated with hGH have not worsened. I share your anxiety regarding Fanconi's anaemia.

M. Sperling, USA: During the symposium I will be presenting data on patients with acromegaly-gigantism [Cutfield et al., abstract 92, Horm Res 1989;31(suppl 1):23]. One of them was 19 months old at the time of diagnosis and, despite multiple therapies, we were eventually able to cure his acromegaly-gigantism at the age of 5. From these studies, we are convinced that a dose of hGH sufficient for any patient will make him grow. Our patient, now 9 years old, persists in having marked insulin resistance as judged by the euglycemic clamp and hyperinsulinemia, perhaps (although I have no evidence for that) reflecting an inexorable effect of hGH at superphysiological doses in the early years of life when there is still self-formation and ongoing organogenesis.

Dr. Brook advocates that we should start treating early, but I raise these issues as a warning that we do not know what doses we are giving that will have permanent rather than reversible effects upon discontinuation of the hGH and, in the long term, may have deleterious effects. Having considered that, I would like to know if anyone has done the opposite of what has been presented today and throughout this symposium. Namely, we are giving increasing doses of hGH: first 3 times a week, then daily and now we are considering the possibility of 3 times daily. However, in our studies, we do not know whether by giving those particular doses to genetically determined individuals we are not inducing a state of acromegaly-gigantism, i.e. by giving what for them are superphysiological doses. Finally, having listened to Dr. Brook and looking at some of the people, including one of our chairmen, I am wondering whether if we give everybody hGH in order to cure some psychological events, we will not perhaps reduce our line of kinship and raise the 3rd percentile for the entire population.

C.G. Brook: I wish to end by saying that size does not seem to always be of importance as I found out in the Israel Museum in the following verses written by King David.

An Apocryphal Psalm
An Hallelujah of David, the son of Jesse (Psalm 151)

(Verse 1)
Smaller was I than my Brothers, and the youngest of the sons of my father.
So he made me Shepherd of his flock, and ruler over his kids.

(Verse 6)
Though they were tall of stature, and handsome by their hair.
The lord God chose, them not.

(Verse 7)
But He sent and took me from behind the flock, and annointed me with Holy Oil.
And He made me leader of His people, and ruler over the sons of his covenant.

Summing-Up of Symposium Update on hGH: Clinical Aspects

A. W. Root, USA

Department of Pediatrics, University of South Florida, College of Medicine, Tampa, Fla., USA

Dr. Preece noted no significant or long-standing adverse effects of treatment upon carbohydrate or lipid metabolism in children with hGH deficiency receiving now what are considered relatively small doses of hGH intermittently. However, he cautioned that the effects of larger doses of hGH administered more frequently for long periods in children without classical hGH deficiency are unknown. Dr. Costin commented that, in her experience with the treatment of such children, increased insulin resistance and hyperinsulinemia were commonly found. Dr. Sperling commented that he had observed severe insulin resistance persisting for many years after removal of a hGH-secreting tumor, when circulating hGH concentrations were < 5 ng/ml. Dr. Mills reported the results of an epidemiological survey of children with hGH deficiency treated with pituitary hGH before the introduction of new preparative procedures in 1976. There are 7 confirmed and 1 probable patient with Creutzfeldt-Jakob disease (CJD) in the USA. In addition, there are 2 hGH-deficient children who have developed CJD who have not been treated with pituitary hGH prepared in the USA. In this survey, there has been no conclusive evidence of an association between hGH therapy and development of leukemia.

Dr. Albertsson-Wikland reviewed data on the spontaneous secretion of hGH, noting in prepubertal children a significant relationship between height and mean 24-hour serum hGH concentration and a significant inverse relationship between the latter measurement and increment in growth when treated with hGH. She commented that it was unlikely that a single test of hGH secretion would be able to distinguish between the child with normal hGH secretion and hGH insufficiency. Dr. Reiter reviewed data on hGH secretion in response to pharmacological stimuli and noted the changing quantitative definitions of an 'adequate' hGH response (from 5 to ⩾ 10 ng/ml) as the availability of hGH for therapy has increased. He discussed data indicating the poor relationship between provoked hGH secretion and linear growth response to hGH. Several members of the audience commented that the most sensitive predictive index of linear growth response to hGH was its inverse relationship to the pretreatment growth velocity.

Dr. Brook reported a survey of European data on the linear growth responses to exogenous hGH in children with a variety of causes of growth retardation not due to hGH insufficiency. In almost every instance, including children with achondroplasia, short-term linear growth rate has increased during hGH therapy. However, in most instances skeletal maturation has also increased; therefore, it is questionable if final adult stature will be increased by such therapy. In boys with constitutional delay in growth and development, oxandrolone has been as efficacious as hGH in accelerating growth in such children and was far less expensive. Dr. Rosenfeld presented data indicating that, after 4–5 years of treatment with hGH or hGH in combination with oxandrolone, many girls with Turner syndrome had heights well above the 50th percentile on the Lyon growth chart and several had heights averaging approximately 150 cm. Analysis of these data in relation to parental height would be of interest. Dr. Job reported that in a double-blind study, hGH significantly increased the short-term growth rate of short children with intrauterine growth retardation.

The major questions which remained unresolved at the conclusion of this symposium were:

(1) Is it possible to differentiate between hGH-sufficient and hGH-insufficient short children? Is it necessary to do so?

(2) What is the optimal manner by which to select short children for treatment with hGH?

(3) What are the long-term growth-promoting effects of hGH in non-hGH-deficient children? Is final adult stature increased? If so, is it of benefit in relation to psychosocial well-being, personality, educational and career achievements?

(4) Can hGH be administered in relatively higher doses for long periods to hGH-sufficient patients without significant adverse side effects related to carbohydrate, lipid and mineral metabolism? Particularly, will long-term hyperinsulinism have adverse effects on atherogenesis, coronary artery disease and so forth?

Horm Res 1990;33:144–151

Multiple Target Antigens in Pre-Type I Diabetes: Implications for Prediction

R. Ziegler, G.S. Eisenbarth

Joslin Diabetes Center, New England Deaconess Hospital, Brigham and Women's Hospital, Harvard Medical School, Boston, Mass., USA

Key Words. Type I diabetes · Insulin autoantibodies · Islet cell antibodies · HLA

Abstract. The prodromal phase of type I diabetes is characterized by the presence of a series of autoantibodies reacting with distinct autoantigens [anti-insulin, anti-64 kDa, islet cell antibodies ('antiganglioside')]. Anti-insulin autoantibodies appear to be unique in that the levels of such antibodies correlate with both the rate of progression to diabetes and age at which type I diabetes develops. With a combination of assays it is now possible to predict type I diabetes, allowing the design of research trials for diabetes prevention.

Introduction

One or more autoantibodies against islet cell antigens can be detected in more than 90% of patients at the onset of type I diabetes, and in some first-degree relatives the presence of autoantibodies precedes the onset of overt diabetes by more than 10 years [1–5]. Autoantibodies and their antigens most intensely studied to date are 'cytoplasmic' islet cell antibodies (ICAs), anti-insulin autoantibodies [6–8] and antibodies to a 64-kilodalton (kDa) protein [9]. Additional antibodies include an antibody reacting in a 'polar' fashion with rat islet tumor (RIN) cells [10], antibodies reacting with the surface antigens of islets or islet tumor cells, antibodies reacting with RIN tumor extracts, as well as a newly described antibody, which reacts with carboxypeptidase H cloned from a rat islet library [11, 12]. A precise knowledge of the nature of these antigens and the prevalence of the autoantibodies could greatly enhance our understanding of the pathogenesis of type I diabetes and aid in the effort to predict the development of diabetes in relatives at risk. In particular, as we will review, different autoantibodies are potentially present in different prediabetics depending on age of diabetes development and/or rate of disease progression.

Islet Cell Antigen

Since the report by Bottazzo et al. [13] that ICAs are a marker for autoimmune β-cell destruction, the predictive value of this marker for type I diabetes has been studied in detail [14–16]. Numerous studies have shown a prevalence of high-titer ICAs in over 60% of new-onset diabetics and approximately 5% of initially normal first-degree relatives of patients with type I diabetes. In population studies of 'normal' schoolchildren, 0.4–0.9% are found to be positive for ICAs [17, 18]. Positivity is measured in Juvenile Diabetes Foundation (JDF) units based on a reference standard serum. In recent workshop comparisons it became clear that there is a large variation both in detection limits in different laboratories (ranging from 5 to 40 or 80 JDF units) and in assay specificity. Prospective studies suggest that only high-titer ICAs (> 40 JDF units) have a positive predictive value greater than 20% with 5 years of prospective evaluation.

The islet cell antigen or antigens responsible for reactivity with frozen sections have not been fully characterized. Nayak et al. [19] found that on frozen sections the autoantigen had the properties of a sialic-acid-containing glycolipid similar to the antigens reacting with antiganglioside monoclonal antibodies A2B5 [20], 3G5 [21] and R2D6 [22] (sensitivity to periodate, restoration with borohydrate, sensitivity to neuraminidase and methanol, and resistance to pronase). In addition, ICA reactivity of sera could be blocked by preincubation with monoganglioside-glycolipid extracts from human pancreas [23]. Islets relative to human pancreas are greatly enriched in a GM_1–GM_2 ganglioside [24]. In collaborative studies with Dotta et al. [25] and Appel et al. [26], the expression of a similar if not the same GM_1–GM_2 ganglioside was metabolically regulable. ICA staining on islets of pancreas from rats suppressed by a RIN insulinoma shows greatly decreased reactivity and in parallel the GM_{2-1} glycolipid is lost [25, 26]. Abrogation of the suppression by removing the transplantable tumor or infusing glucose rapidly restores ICA reactivity of the pancreas and expression of the GM_{2-1} ganglioside. It is thought that this resting of the β-cells is responsible for the diminished expression of the antigen, which rebounds readily only 1 day after the removal of the tumor and the normalization of glycemia [25, 26]. Gillard et al. [27] have isolated a different ganglioside from RIN tumors (GT3) and directly demonstrated reactivity of type I diabetic sera with this ganglioside.

Insulin

Palmer et al. [7] first described anti-insulin autoantibodies in newly diagnosed diabetics before insulin treatment. Since then multiple studies have been performed by different groups studying new-onset diabetics as well as first-degree relatives of patients with type I diabetes. Two different assay formats are being used, radioimmunoassay (RIA) and enzyme-linked immunosorbent assay (ELISA). In recent workshop comparisons at the Immunology and Diabetes Workshops in Auckland it became apparent that the two assays do not measure the same antibodies. ELISAs appear to have a high rate of false-positive results, with up to 20% of first-degree relatives measuring as positive. Because it is known that only approximately 5% of these relatives will ever develop type I diabetes, the predictive value of this assay format is questionable. We utilize a competitive anti-insulin RIA with polyethylene glycol precipitation [28] modified after the method from Palmer et al. [7]. The results are expressed as nanounits per milliliter precipitated insulin and our upper limit of normal is 39 nU/ml (3 SD above normal mean of 74 population controls with no family history of diabetes). Approximately 4% of first-degree relatives exceed the upper limit of normal in our RIA for competitive insulin autoantibodies (CIAAs). Forty-five percent of ICA-positive relatives are CIAA-positive [29]. There is a unique inverse relationship between age at which type I diabetes develops and the level of CIAAs. Nearly 100% of children developing diabetes under the age of 5 years but only approximately 20% over the age of 15 have CIAA levels above 39 nU/ml at the time of diagnosis of overt diabetes, with the CIAA levels for the young children often exceeding 2,000 nU/ml [30]. The level of CIAAs also correlates with the long-term persistence and constancy of the antibodies and the rate of progression to overt diabetes in ICA-positive relatives [31, 32]. Of 23 ICA-negative (or low-titer) relatives with CIAA levels between 40 and 80 nU/ml, only 7 (30%) were positive on subsequent sample determination compared to 19/22 (86%) in the group with CIAA >80 nU/ml. In ICA-positive relatives the influence of the CIAA level on persistence is not apparent as all such CIAA-positive relatives almost always remain positive until diabetes develops. The 4 ICA-negative/CIAA-positive relatives in our series who developed diabetes all had CIAA levels >80 nU/ml [32]. The level of CIAAs in relatives is remarkably constant over years of follow-up for each individual and we see no trend in levels as diabetes approaches. Recently, we found an association of the level of CIAAs with HLA-DR4 in autoantibody-positive first-degree relatives [33]. Twenty-four percent (25/104) of relatives with at least 1 DR4 allele exceeded our upper limit of normal (39 nU/ml) compared to 7% (10/135) of those with no DR4 allele. Even more striking was the relationship with a level of CIAAs >80 nU/ml, where 15% (16/104) of the relatives with at least 1 DR4 allele had CIAA levels >80 nU/ml compared to 1% (2/135) of those with no DR4 allele. All but 1 of 15 relatives with the highest CIAA valus were in the DR4 group. To define the region of the insulin molecule with which patients' autoantibodies react, we have evaluated the reactivity of a series of insulins and insulin analogues. All antibodies were of high affinity (10^{10} liters/mol and low capacity (10^{-12} mol/liters. Guinea pig and fish insulin did not compete in our assay, while human, porcine, bovine, rat, chicken and sheep insulin reacted but varied in the concentration of insulin required for half-maximal inhibition [34]. Additionally, various insulin analogues with specified deletions or substitutions were used and we now hypothesize that the immunodominant region is residues B1–B4 and A8–A13. All studied sera of relatives behaved in a similar pattern, suggesting the sample immunodominant epitope for all [35].

It is not yet known if insulin is a primary target and plays a crucial role in the pathogenesis of type I diabetes. Circumstantial evidence suggests that an immune response to insulin could at least enhance the immunologic destruction of β-cells. Larsson et al. [36] demonstrated that insulin is expressed on the surface of intact β-cells accessible to antibodies and thus may be a target for antibodies or T lymphocytes. The correlation between the levels of anti-insulin autoantibodies and the rate of progression to diabetes, the correlation with the age at which diabetes develops and the correlation with DR4 all suggest that the immune response to insulin is part of the pathogenesis of diabetes. It is noteworthy that neither antibodies to the 64-kDa protein nor ICAs correlate with the age at which type I diabetes develops.

64-kDa Protein

Baekkeskov et al. [9] and Christie et al. [37] found antibodies to a 64-kDa protein in newly diagnosed children with type I diabetes by immunoprecipitation of ^{35}S-labeled human islet lysates. Approximately 70% of new-onset diabetic children are positive for this antibody [9, 37]. Furthermore, it has been shown that these antibodies can precede the clinical onset of diabetes and can be present in first-degree relatives of type I diabetics with impaired β-cell function [38, 39]. The location of the antigen in the β-cell and the characterization of its molecular structure have not yet been elucidated. Colman et al. [40] have provided evidence that the 64-kDa antigen is not expressed on the surface of cultured human islet cells. The recent utilization of rat islets has made the isolation process easier but the assay remains very laborious and not yet suited for large screening programs [37].

Additional Antigens and Autoantibodies

A number of other autoantibodies in type I diabetics and first-degree relatives of diabetic patients have been described. These include autoantibodies reacting with antigens at the cell surface of RIN tumor cells [10] or in a 'polar' fashion with RIN tumor cells [11] and with carboxypeptidase H cloned from a rat islet library [12]. Apart from antigens of pancreatic tissue or from pancreatic-derived tumors, autoantibodies against antigens of nonpancreatic, but common neuroendocrine [41], origin can be detected in type I diabetics, e.g. antithyroid microsomal [42, 43], antiadrenal medullary [44, 45], antisympathetic ganglion [46], antigastric parietal cell [42, 43] and anticarcinoid tumor [47] antibodies. Furthermore, autoantibodies against proinsulin [48], insulin receptors [49, 50] and immunoglobulins of different species [51] have been demonstrated in new-onset diabetic patients.

Of the 'pancreas tissue'-derived antigens, only the sequence of carboxypeptidase H is known by virtue of the method used to detect the autoantibodies. The antigen was detected by screening with prediabetic sera on a rat islet expression library. Through cloning and sequencing the identified positive colony the sequence of the autoantigen could be derived [12]. Further investigations using a human islet library are underway to eventually synthesize the protein and then to be able to screen larger cohorts with immunoprecipitation assays utilizing labeled human carboxypeptidase H.

The cell surface antigens on RIN tumor cells have not been characterized to date. RIN tumor cells can lose their antigenicity in culture and this was attributed to a loss of ganglioside expression of the cells. This suggests a glycolipid nature of some of the antigens [11]. For the 'polar' antigen, so called because of the appearance of the staining on the pole of RIN cells, (antibodies against it can be detected in approximately 20% of new-onset diabetics), a protein and not a glycolipid is thought to be the target molecule [10].

Prediction of Type I Diabetes

With the occurrence of autoantibodies in first-degree relatives, the prediction of the risk to develop type I diabetes and the probable time to onset of the disease in first-degree relatives is a current challenge. The ultimate reason to identify a person at risk for type I diabetes would be the availability of an effective and safe intervention in the prodromal stage of the disease to prevent diabetes. Apart from autoantibodies detected before onset of type I diabetes, diminished first-phase insulin response to the intravenous glucose tolerance test (IVGTT) is frequently found prior to overt diabetes [52]. The correlation of the level of anti-insulin autoantibodies and time to overt diabetes suggested that insulin autoantibodies reflect the rate at which β-cell destruction occurs while the loss of first-phase insulin secretion of IVGTT the degree of β-cell loss.

In figure 1 the level of anti-insulin autoantibodies in first-degree relatives is plotted versus years of follow-up to diabetes or continuing nondiabetic follow-up. Relatives with the highest levels of insulin autoantibodies rarely remain nondiabetic for more than 3 years. In those relatives who had an insulin release after 1 + 3 min $< 48\ \mu U/ml$ (below the first percentile in our insulin assay) on their first IVGTT, the level of CIAAs was an independent determinant of the rate of progression to overt type I diabetes, as seen in years of follow-up to disease onset (fig. 1b). In those relatives with a 1 + 3 min insulin release $> 48\ \mu U/ml$, the correlation between the CIAA level and years of follow-up to overt diabetes was stronger with an r value = 0.92, $p < 0.01$ (fig. 1c). This demonstrates the independent nature of this influence of the level of CIAAs on the rate of progression to overt diabetes and that in relatives with first-phase insulin release greater than the first percentile the correlation is linear. Jackson et al. [53] have used these parameters in ICA-positive first-degree relatives to develop a linear formula to predict the approximate onset of overt diabetes. The formula developed with 14 individuals who became diabetic on follow-up had a correlation coefficient of r = 0.76, $p < 0.001$. Further studies on a larger number of relatives with this formula supported these results [53]. This linear formula implies a constant level of anti-insulin autoantibodies or at least levels which do not systematically change with time. We have found that CIAA values are indeed relatively constant over long periods of time in ICA-positive and -negative relatives without a rise of CIAA levels before onset of the disease (fig. 2, a typical subset of relatives we have followed) [32]. The values of each sequential sample for several individuals over time of follow-up are shown in figure 2 and demonstrate that, although the absolute values are not constant, they vary in ranges characteristic for each relative over long periods of time. Independent from the results of IVGTT, the level of anti-insulin autoantibodies itself can be used to calculate the risk for development of type I diabetes by life table analysis [29]. In ICA-positive relatives with CIAA levels > 150 nU/ml, 100% were predicted to become diabetic within 5 years, compared to approximately 50% of those with levels below 150 nU/ml or negative for CIAA. Similar analyses for ICA-negative relatives, although with fewer subjects, showed that none in the group of ICA-negative/CIAA-positive relatives with values between 40 and 80 nU/ml became diabetic in contrast to approximately 40% of relatives with CIAA levels > 80 nU/ml by life table analysis [32].

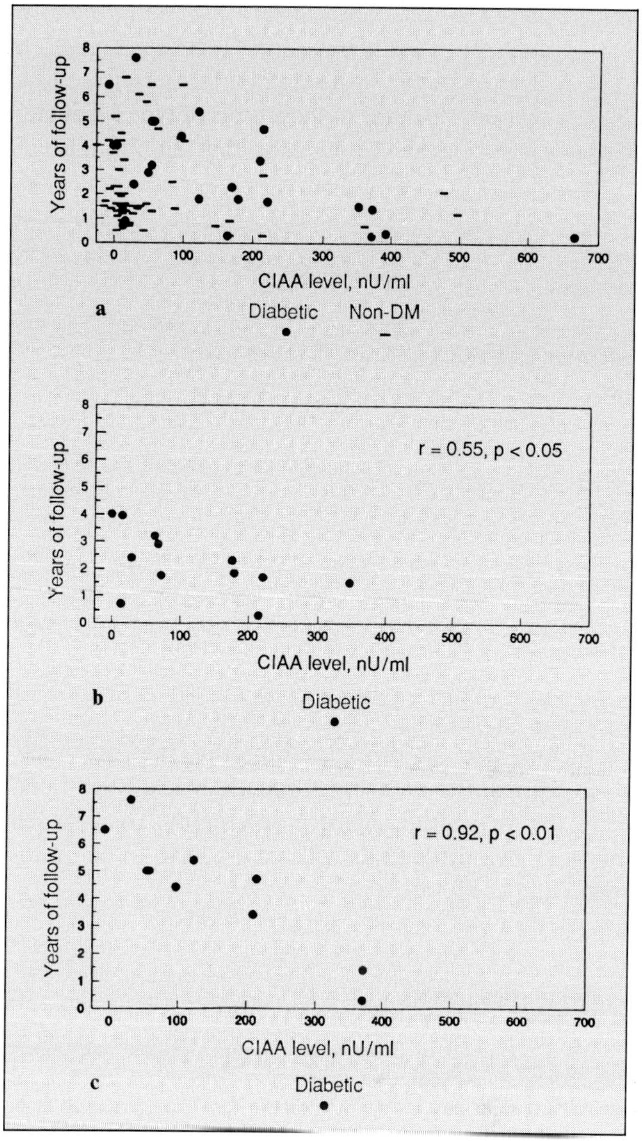

Fig. 1. CIAA levels versus years of follow-up in ICA-positive first-degree relatives. **a** All followed relatives are shown subdivided by their subsequent development of overt type I diabetes, regardless of their IVGTT result: IVGTT < or > 48 μU/ml or no IVGTT. **b** Relatives with a 1 + 3 min insulin release on their first IVGTT < 48 μU/ml (below the first percentile), who developed diabetes on follow-up. **c** Relatives with an IVGTT > 48 μU/ml and subsequent development of overt type I diabetes.

In summary, we believe that with a combination of research autoantibody assays and metabolic tests it is possible to identify first-degree relatives with a very high risk for type I diabetes. This information should allow the design of intervention trials in the prodromal phase of type I diabetes. Currently, two such trials are approved at the Joslin Diabetes Center, one utilizing insu-

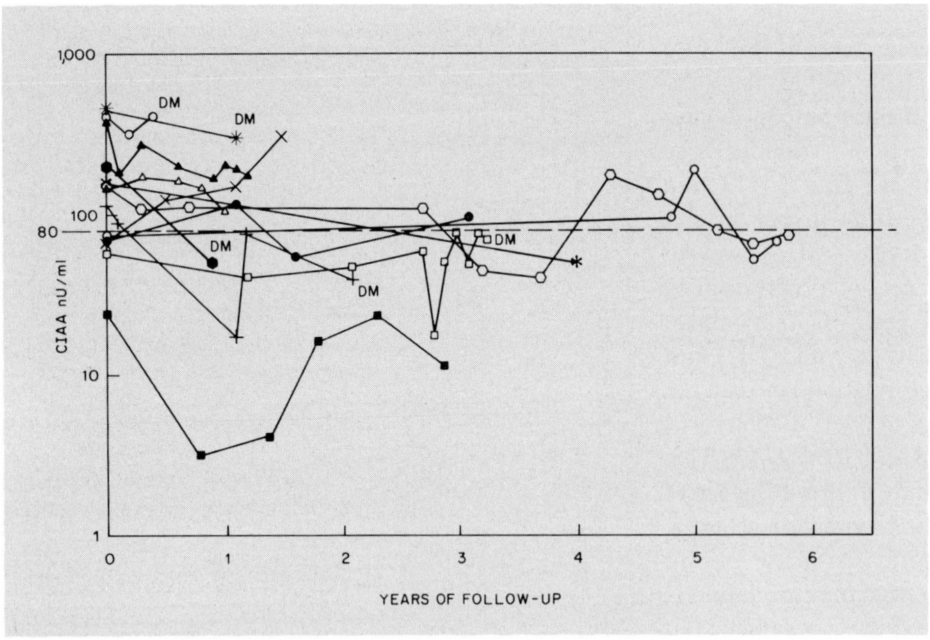

Fig. 2. CIAA levels over time of follow-up for 13 first-degree relatives. DM indicates development of type I diabetes at this point of follow-up.

lin therapy and the other azathioprine. It is our hope, given the realities of current management of type I diabetes, that prevention can become a reality. To approach this goal, definitive trials will almost have to be multicenter and cooperative.

Acknowledgments

We are grateful to Irene Reske, Maura Tautkus and Albert Rabizadeh for excellent technical support, Theresa Smith, RN, for her nursing skills and Patricia Cronin-Sevigny for secretarial help. The NIH Diabetes and Endocrinology Research Center grant (DK36836) with Clinical and Radioligand Cores were essential to this study. This work was supported by grant DK32083 from the National Institute of Health and a grant from the Eleanor Dana Charitable Trust, the JDF and the American Diabetes Association. R.Z. is supported by the 'Deutsche Forschungsgemeinschaft'.

References

1 Rossini AA, Mordes JP, Handler ES: Speculations on etiology of diabetes mellitus: Tumbler hypothesis. Diabetes 1988;37:257–261.
2 Eisenbarth GS: Type I diabetes mellitus: A chronic autoimmune disease. N Engl J Med 1986;314:1360–1368.
3 Segall H: HLA and genetics of IDDM: Holism vs reductionism. Diabetes 1988;37:1005–1008.
4 Landin-Olsson M, Karlsson A, Dahlquist G, Blom L, Lernmark A, Sundkvist G, et al: Islet cell and other organ-specific autoantibodies in all children developing type I (insulin-dependent) diabetes mellitus in Sweden during one year and in matched control children. Diabetologia 1989;32:387–395.
5 Kolb H, Dannehl K, Grüneklee D, Zielasek J, Bertrams J, Hübinger A, Gries FA: Prospective analysis of islet cell antibodies in children with type I (insulin-dependent) diabetes. Diabetologia 1988;31:189–194.
6 Gorsuch AN, Spencer KM, Lister J, McNally JM, Dean BM, Bottazzo GF, Cudworth AG: Evidence for a long prediabetic period in type I (insulin dependent) diabetes mellitus. Lancet 1981;ii:1363–1365.
7 Palmer JP, Asplin CM, Clemons P, Lyon K, Iatpati O, Raghu P, Paquette TL: Insulin antibodies in insulin dependent diabetes before insulin treatment. Science 1983;222:1337–1339.
8 Soeldner JS, Tuttleman M, Srikanta S, Ganda OP, Eisenbarth GS: Insulin dependent diabetes mellitus and initiation of autoimmunity: Islet cell autoantibodies, insulin autoantibodies and β cell failure. N Engl J Med 1985;313:893–894.
9 Baekkeskov S, Nielsen JH, Marner B, Bilde T, Ludvigsson J, Lernmark A: Autoantibodies in newly diagnosed diabetic children immunoprecipitate human pancreatic islet cell proteins. Nature 1982;298:167–169.
10 Dotta F, Bonner-Weir S, Cahill C, Appel M, Nayak RC, Eisenbarth GS: A new diabetes-associated antigen: Evidence that it is a protein and not a glycolipid. Diabetologia 1988;31:486A.
11 Thomas JW, Virta VJ, Nell LJ: Cell surface antigens on rat islet tumors. J Immunol 1987;138:2896–2901.
12 Russo E, Castano L, Lipes M, Zhou LJ, Eisenbarth GS: Identification and cloning of granule autoantigen (carboxypeptidase H) associated with type I diabetes. Submitted.
13 Bottazzo GF, Gorsuch AN, Dean BM, Cudworth AG, Donaich D: Complement-fixing islet cell antibodies in type I diabetes: Possible monitors of active β cell damage. Lancet 1980;i:668–672.
14 Gorsuch AN, Spencer KM, Lister J, Wolf E, Bottazzo GF, Cud-

worth AG: Can future type I diabetes be predicted? A study in families of affected children. Diabetes 1982;31:862–866.
15 Srikanta S, Ganda OP, Rabizadeh A, Soeldner JS, Eisenbarth GS: First degree relatives of patients with type I diabetes mellitus: Islet cell antibodies and abnormal insulin secretion. N Engl J Med 1985;313:461–464.
16 Riley WJ, Spilar RP, Waltz J, Brody B: Predictive value of islet cell antibodies (ICA): Six year experience. Diabetes 1984; 33(suppl 1):44A.
17 MacLaren NK, Horne G, Spillar RP, Barbour H, Harrison D, Duncan J: Islet cell autoantibodies (ICA) in U.S. school children. Diabetes 1985;34(suppl 1):84A.
18 Bergua M, Sole J, Marion G, Perez MC, Recasens A, Fernandez J, Casamitjana R, Gomis R: Prevalence of islet cell antibodies, insulin antibodies and hyperglycemia in 2291 school children. Diabetologia 1987;30:724–726.
19 Nayak RC, Omar MAK, Rabizadeh A, Srikanta S, Eisenbarth GS: 'Cytoplasmic' islet cell antibodies: Evidence that the target antigen is a sialoglycoconjugate. Diabetes 1985;34:617–619.
20 Eisenbarth GS, Shimizu K, Bowring MA, Wells S: Expression of receptors for tetanus toxin and monoclonal antibody A2B5 by pancreatic islet cells. Proc Natl Acad Sci USA 1982;79:5066–5070.
21 Powers ACA, Rabizadeh A, Akenson R, Eisenbarth GS: Characterization of monclonal antibody 3G5 and utilization of this antibody to immobilize pancreatic islet cell gangliosides in a solid phase radioassay. Endocrinology 1984;114:1338–1343.
22 Alejandro R, Shienvold L, Hajek SAV, Pierce M, Paul R, Mintz DH: A ganglioside antigen on the rat pancreatic B cell surface identified by monoclonal antibody R2D6. J Clin Invest 1984;74: 25–38.
23 Colman PG, Nayak RC, Campbell IL, Eisenbarth GS: Binding of cytoplasmic islet cell antibodies is blocked by human pancreatic glycolipid extracts. Diabetes 1988;37:645–652.
24 Dotta F, Colman PG, Nayak RC, Lombardi D, Scharp DW, Andreani D, Pontieri GM, Di Mario U, Lenti L, Eisenbarth GS: Ganlioside expression in human pancreatic islets. Diabetes 1989;38:1478–1483.
25 Dotta F, Ziegler R, O'Neill JJ, Nayak RC, Eisenbarth GS, Appel MC: Islet autoimmunity: Identification and initial characterization of a metabolically regulable pancreatic ganglioside. Diabetologia 1989;32:483A.
26 Appel M, Dotta F, O'Neil JJ, Eisenbarth GS: B-cell activity regulates the expression of islet antigenic determinants. Diabetologia 1989;32:461A.
27 Gillard BK, Thomas JW, Nell LJ, Marcus DM: Antibodies against ganglioside GT3 in the sera of patients with type I diabetes mellitus. J Immunol 1989;142:3826–3832.
28 Vardi P, Dib SA, Tuttleman M, Connelly JE, Grinbergs M, Rabizadeh A, Riley WJ, MacLaren NK, Eisenbarth GS, Soeldner JS: Competitive insulin autoantibody assay: Prospective evaluation of subjects at high risk for development of type I diabetes mellitus. Diabetes 1987;36:1286–1291.
29 Ziegler AG, Ziegler R, Vardi P, Jackson RA, Soeldner JS, Eisenbarth GS: Life-table analysis of progression to diabetes of anti-insulin autoantibody-positive relatives of individuals with type I diabetes. Diabetes 1989;38:1320–1325.
30 Vardi P, Ziegler AG, Mathews JH, Dib S, Keller RJ, Ricker AT, Wolfsdorf JI, Herskowitz RD, Rabizadeh A, Eisenbarth GS, Soeldner JS: Concentration of insulin autoantibodies at onset of type I diabetes: Inverse log-linear correlation with age. Diabetes Care 1988;11:736–739.
31 Jackson RA, Soeldner JS, Eisenbarth GS: Testing the dual parameter model for predicting time to diabetes. Clin Res 1989; 37:571A.
32 Ziegler R, Ziegler AG, Jackson RA, Eisenbarth GS: The level of 'positive' competitive insulin autoantibodies correlates with development of diabetes mellitus and constancy of the autoantibodies. Diabetologia 1989;32:560A.
33 Ziegler R, Alper CA, Awdeh ZL, Castano L, Yunis EJ, Brink SJ, Jackson RA, Eisenbarth GS: Specific association of HLA-DR4 with increased prevalence and level of anti-insulin autoantibodies in first degree relatives of patients with type I diabetes. Submitted.
34 Castano L, Ziegler AG, Ziegler R, Russo E, Jackson RA: Quantitation of competitive insulin autoantibodies (CIAA). Diabetes 1989;38(suppl 2):185A.
35 Castano L, Ziegler AG, Ziegler R, Shoelson S, Eisenbarth GS: Anti-insulin autoantibodies: Recognition of a limited epitope distinct from the receptor binding domain. Submitted.
36 Larsson LI, Nielsen JH, Hutton JC, Madsen OD: Pancreatic hormones are expressed on the surfaces on human and rat islet cells through exocytotoxic sites. Eur J Cell Biol 1989;48:45–51.
37 Christie M, Landin-Olsson M, Sundkvist G, Dahlquist G, Lernmark A, Baekkeskov S: Antibodies to a Mr-64000 islet cell protein in Swedish children with newly diagnosed type I (insulin-dependent) diabetes. Diabetologia 1988;31:597–602.
38 Baekkeskov S, Landin M, Kristensen JK, Srikanta S, Bruining GJ, Mandrup-Poulsen T, de Beaufort C, Soeldner JS, Eisenbarth GS, Lindgren F, Sundquist G, Lernmark A: Antibodies to a 64000 Mr islet cell antigen precede the clinical onset of insulin-dependent diabetes. J Clin Invest 1987;79:926–934.
39 Atkinson MA, MacLaren NK, Riley WJ, Scharp DW, Holmes L: Mr 64000 autoantibodies (64KA) predict insulin dependent diabetes (IDD). Diabetes 1988;37(suppl 1):98A.
40 Colman PG, Campbell IL, Kay TWH, Harrison LC: 64000-Mr autoantigen in type I diabetes: Evidence against its surface location on human islets. Diabetes 1987;36:1432–1440.
41 LeDourain NM. On the origin of pancreatic endocrine cells. Cell 1988;53:167–171.
42 Goldstein E, Drash A, Gibbs J, Blizzard RM: Diabetes mellitus: The incidence of circulating antibodies against thyroid, gastric, and adrenal tissue. J Pediatr 1970;77:304–306.
43 Nerup J, Binder C: Thyroid, gastric and adrenal auto-immunity in diabetes mellitus. Acta Endocrinol 1973;72:279–286.
44 Schopfer K, Matter L, Tenschert R: Anti-glucagon cell and anti-adrenal medulla cell antibodies in islet cell autoantibody positive diabetic children. N Engl J Med 1984;310:1536–1537.
45 Brown FM, Kamalesh M, Sesh Adri MN, Rabinowe SL: Anti-adrenal medullary antibodies in IDDM subjects and subjects at high risk of developing IDDM. Diabetes Care 1987;11:30–33.
46 Rabinowe SL, Brown FM, Watts M, Kadrofske MM, Vinik AI: Anti-sympathetic ganglia antibodies and postural blood pressure in IDDM subjects of varying duration and patients at high risk of developing IDDM. Diabetes Care 1989;12:1–6.
47 Miettinen A, Holthöfer H, Kontiainen S, Miettinen M, Andersson LC: Antibodies against gastrointestinal carcinoid tumors in IDDM. Diabetes 1989;38:667–669.
48 Kuglin B, Gries FA, Kolb H: Evidence of IgG autoantibodies

against human proinsulin in patients with IDDM before insulin treatment. Diabetes 1988;37:130–132.
49 Ludwig SM, Faiman C, Dean HJ: Insulin and insulin-receptor autoantibodies in children with newly diagnosed IDDM before insulin therapy. Diabetes 1987;36:420–425.
50 Maron R, Elias D, de Jongh BM, Bruining GJ, van Rood JJ, Schechter Y, Cohen IR: Autoantibodies to the insulin receptor in juvenile onset insulin-dependent diabetes. Nature 1983;303: 817–818.
51 Di Mario U, Dotta F, Crisa L, Anastasi E, Andreani D, Dib SA, Eisenbarth GS: Circulating anti-immunoglobulin antibodies in recent-onset type I diabetic patients. Diabetes 1988;37:462–466.
52 Srikanta S, Ganda OP, Gleason RE, Jackson RA, Soeldner JS, Eisenbarth GS: Pre-type I diabetes: Linear loss of β cell response to intravenous glucose. Diabetes 1984;33:717–720.
53 Jackson RA, Vardi P, Herskowitz RD, Soeldner JS, Eisenbarth GS: Dual parameter linear model for prediction of onset of type I diabetes in islet cell antibody positive relatives. Clin Res 1988;36:484A.

Dr. G.S. Eisenbarth
Joslin Diabetes Center
One Joslin Place
Boston, MA 02215 (USA)

Discussion

M.A. Sperling, USA: I would like to express a word of caution. We have a paper in the poster session in which we had the opportunity to prospectively study approximately 50 siblings of patients with classical type I diabetes. We present a case of a young patient in whom the ICA level was positive and first-phase insulin secretion normal so that, based on the predictive formula that the Joslin group has published, this child would not develop diabetes for 10 or more years. However, insulin-dependent diabetes mellitus developed within 8 months. How do you reconcile this?

G.S. Eisenbarth, USA: Your exception is of interest and we would like to measure anti-insulin autoantibodies in this individual.

P. Saenger, USA: Does the measurement of CIAA levels help to explain the observed increase in the incidence of type I diabetes mellitus? Have CIAA levels been measured over time?

G.S. Eisenbarth: No data for CIAA levels over time are available.

G.J. Bruining, The Netherlands: The data you presented are essentially retrospctive. These are used as a model for prospective prediction. Can this be called a 'linear model' for prediction?

G.S. Eisenbarth: The data are now prospective with 10 individuals progressing to diabetes after the first 14 upon whom the model was based.

J. Nerup, Denmark: You put a lot of emphasis on CIAA in your equation. This prompts two questions in relation to ICA positivity: (a) Does your model hold true in ICA-negative individuals? (b) Do you see insulin as the pathogenetically primary antigen?

G.S. Eisenbarth: (a) It is not yet known if the formula will work in ICA-negative relatives who are CIAA-positive. (b) I believe it is likely that insulin is a primary target antigen, but testing this hypothesis will take studies involving prevention of diabetes by preventing immune responses to insulin. My answer is based on the rate correlation and expression of insulin on the cell surface.

J. Nerup: You showed a slide of predicted time to diabetes against observed time to diabetes. Clearly there was a correlation but no r value or r^2 value was given. This may be interesting to the audience. An r value of > 0.9 would allow some individual predictability whereas lower values may not.

G.S. Eisenbarth: r values for correlation with anti-insulin antibodies and time to diabetes are as high as 0.92. One can use the prediction for the extremes and thus can assign a 90% risk of diabetes within 3 years for selected relatives.

P. Czernichow, France: Do you have any information regarding the natural history of ICA-positive patients? Do your healthy patients become positive and are there any clinical events associated with the conversion from negativity to positivity?

G.S. Eisenbarth: We have only observed 2 patients who converted to positivity. Almost all our patients when they are screened as relatives are already positive on the first serum sample. The 2 people that we have seen, one converted insulin autoantibodies between 1 and 4 years of age and the second was a twin that Dr. Sullivan followed, who converted at the age of 31 years, 5 years before diabetes. We looked for viruses, not viral autoantibodies, but we have no information as to what triggers what, then reverts to a steady level of antibodies.

A.L. Drash, USA: You know that for a long time we have been concerned about the fact that your formula does not take in the possibility of age variation. You showed us that there is a major age variation in terms of antibody to insulin and the fact that entered your formula causes me concern regarding the insulin response. I think we probably have data from our own work that would indicate that, as in the oral glucose tolerance test, the IVGTT also has age relationships. Ingeborg Deschamps showed us last week at the ISGD Meeting some work on the insulin response to an IVGTT by age and it is clear that the median and the lower third percentile are quite different depending upon age and somehow that has to be entered into the formula for accuracy.

G.S. Eisenbarth: I agree fully with the statement that around puberty the first-stage insulin release increases compared to especially very young children. I think it only changes around 2- to 2.5-fold, thus there is a potential for deriving a better formula, e.g. percentiles based on age for first-stage insulin response given the subset of data that are available, but it will be a long time before one has that. We actually did try to put age into the formula in the linear regression and it has a remarkably small effect once you know the level of insulin autoantibodies.

H.B. Mortensen, Denmark: Insulin autoantibodies were present in high titers in children with onset of diabetes below 5 years of age. In contrast, lower levels were observed in the age groups over 5 years. Do you think this suggests a heterogeneity within type I diabetes?

G.S. Eisenbarth: I believe there is heterogeneity primarily in the rate of progression with younger children progressing faster and developing diabetes earlier.

I. Deschamps, France: I would like to comment on our experience regarding insulin antibodies. There is a great individual variability of insulin antibodies but perhaps that is a question of the technique used. Have you seen individuals who had very high titers of antibodies and a few months later had no antibodies? In a prospective study we have followed more than 200 relatives of diabetic children and those who developed diabetes were found to have insulin antibodies only a few months before onset of the disease. I think that limits the use of your model. Anyhow you showed insulin antibodies as seen only in very young children. What model do you use in children over the age of 5?

G.S. Eisenbarth: Your statements are dependent up on the technique used.

I. Deschamps: The dose determinations were made by Prof. Bach's laboratory in Paris.

G.S. Eisenbarth: His assay is less sensitive than ours and I think that is part of the difference in terms of assay variability. I had 2 other slides that I did not have time to show. We have seen 3 relatives who had insulin autoantibodies, 100 and 200 nU/ml, who on the next sample had no autoantibodies. That is 3 out of 30 patients. Once insulin autoantibodies have appeared twice, we have never seen them disappear. In patients who have insulin autoantibodies between 40 and 80 nU/ml (just barely above our upper limit of normal), 90% on the second sample will not have detectable insulin autoantibodies. Therefore, I think in the low range there is a lot of variability but not much in the high range. As to the idea that insulin autoantibodies appear shortly before the onset of type I diabetes, we do not have the data to support that. I showed you 8 different patients who had insulin autoantibodies over 2- to 5-year time periods and we have seen 1 patient lose his insulin autoantibodies right before type I diabetes and no clear upward trend, so the antibodies, at least for this RIA, are there for long periods of time.

Horm Res 1990;33:152–158

0301–0163/90/0334–0152$2.75/0

Induction and Pathophysiology of Remission of Insulin-Dependent Diabetes Mellitus during Administration of Ciclosporin

J. Dupre, I. Hramiak, J.L. Mahon, C.R. Stiller (for the London Diabetes Study Group)

University Hospital, London, Ont., Canada

Key Words. Insulin-dependent diabetes · Remission · Ciclosporin · Insulin-connecting peptide · Insulin resistance

Abstract. Ciclosporin-induced noninsulin-receiving remissions in insulin-dependent diabetes mellitus are associated with enhancement of beta-cell function. Patients in remission show virtually no insulin responses to parenteral glucose, contrasting with substantial responses to mixed meals, indicating effects of nonglucose nutrients and/or enteroinsular mechanisms. Remission is associated with normal insulin sensitivity; loss of insulin sensitivity can lead to relapse. The risks of the treatment call for studies with lower doses, and it is argued that a continuing search for immunomodulatory interventions that enhance and preserve the clinical remission-phase characteristics should be pursued.

Introduction

The rationale for experimental immunotherapy in overt insulin-dependent diabetes mellitus (IDDM) was that suppression of an autoimmune attack on beta-cells might preserve or enhance beta-cell function and result in clinical 'remission' of the disease [1]. The potential for such presumably beneficial effects was suggested by studies of secretion of insulin in IDDM, showing that residual beta-cell function commonly improves for a period of weeks following diagnosis and initiation of treatment with insulin, to an optimum level at 3 or 4 months after onset [2]. At this time glycemic control also optimizes, with insulin dosage at a nadir, suggesting a role for the improved beta-cell function in such partial clinical remission.

Definition of Remission in IDDM

True remission of IDDM with disappearance of all features whereby the disease can be recognized is virtually unknown. However, the diminishing requirement for insulin with maintained metabolic control that occurs in the early months after diagnosis characterizes a 'clinical remission phase' of the disease which has been generally recognized and often defined in terms of insulin dosage. In studies of remission-phase diabetes, the conditions under which reduction of insulin dose or withdrawal of the therapy are employed to define remission are clearly important. Strict normoglycemia has usually not been required in recognition of noninsulin-receiving (NIR) remissions in studies of this kind. This is based on the view that maintenance of target glycemic control at levels below those diagnostic of diabetes mellitus in the absence of insulin therapy represents a reasonable operating definition. The response to experimental therapies can also be defined in terms of beta-cell function, and such effects could be employed in simple or compound definitions of remission. In the present account, emphasis is placed on the results of studies with ciclosporin as the immunomodulatory agent, because to date this is the only one that has exhibited significant effects on clinical remissions that have also been associated with unequivocal enhancement of beta-cell function [3].

Effects of Ciclosporin in the Clinical Remission Phase of IDDM

The first open (uncontrolled) study examined the effects of ciclosporin, given in an initial dose of 10 mg/kg/day with adjustment to maintain serum trough levels in the range of 150–200 ng/ml (polyclonal radioimmunoassay; RIA) in patients with clinically diagnosed IDDM of up to 11 months duration [1]. The findings indicated apparent induction of remission to a NIR state with maintained target control of glycemia (< 7.8 mmol/l before meals) in an unexpectedly high proportion of patients. This was associated with an early increase in fasting C peptide levels in the plasma, which were maintained at a near-normal plateau through 12 months. The NIR remission rate of approximately 40% at 1 year in patients who entered within 8 weeks of starting therapy was strikingly higher than that recorded in historical experience with conventional insulin therapy. Moreover, the maintenance of stable levels of C peptide in the plasma through the second half of the year of study in patients with 'recent-onset' diabetes (< 8 weeks insulin therapy) was apparently unprecedented.

Continued experience in the Canadian open study, in which the great majority of the 81 patients were aged 9 years or more at the time of clinical diagnosis, was consistent with the initial findings, and led to a number of inferences regarding the characteristics of patients more likely to respond to ciclosporin [4]. Within the group who had been treated with insulin for no more than 8 weeks at entry, there was a significantly higher chance of development of NIR remission in those with insulin therapy not exceeding 3 weeks in duration at entry. Moreover, the chances of remission were greater in patients aged 16 years or more at entry, by comparison with younger patients. Yet lower rates of remission were observed in a further group of younger patients, ranging in age from 22 months to 8 years at entry [5]. The clinical outcome did not correlate with the basal or stimulated C peptide level in the plasma at entry, or with the presence or absence of islet cell antibodies in the blood. The Canadian open studies also yielded the important information that interruption of treatment with ciclosporin at intervals of from 12 to more than 30 months from entry was likely to be followed by relapse within a few weeks. An open study with ciclosporin in France yielded similar results in terms of remissions defined by the insulin treatment status [6]. The anticipated clinical and biochemical adverse effects of ciclosporin observed in these studies were not different from those documented in other experience with the drug, and were fully reversible on interruption of therapy.

Randomized Control Trials with Ciclosporin

The first report of a randomized control trial of ciclosporin in IDDM came from a multicenter study group in France [7]. In this trial, 122 patients were randomized to treatment with ciclosporin or placebo, with continued administration of insulin, but with attempted minimization of the dose as consistent with maintenance of target control of glycemia. The patients and those managing the diabetes were blinded to the treatments. Ciclosporin was administered in predetermined dosage, without adjustment on the basis of blood levels of the drug. Initially the dose was 7.5 mg/kg/day; this was increased to 10 mg/kg after 3 months if the insulin dose exceeded 50% of that required for metabolic control at entry. So-called 'complete remission' was identified when blood glucose levels were maintained < 7.8 mmol/l before meals and < 11.1 mmol/l after meals, with glycohemoglobin levels ≤ 7.5% (upper limit of normal 5.8%), without administration of insulin. Partial remission was recognized when target metabolic control was achieved with insulin dosage < 0.25 U/kg/day. After 6 months, administraton of ciclosporin was discontinued in patients who showed no evidence of remission. Patients in remission continued to take ciclosporin, but the dose was decreased by 20% monthly, to reach 5 mg/kg/day before 9 months. At the time of reporting, data from 106 patients were available. The complete-remission rates at 9 months were 24% in the ciclosporin-treated group and 6% in the placebo-treated group ($p < 0.01$). When the findings were examined in relation to the observed blood levels of ciclosporin, it was found that a substantial number of patients had maintained levels in the range considered to be subtherapeutic on the basis of experience in organ transplantation (i.e. < 300 ng/ml whole-blood trough levels, polyclonal RIA). At 6 months the complete-remission rate was 37.5%, and the complete-plus-partial remission rate was 65.5%, in those patients in whom the ciclosporin trough level had been maintained at 300 ng/ml or more through the first 3 months, compared to 21 and 32%, respectively, in placebo-treated patients. This study demonstrated that ciclosporin can induce remissions of type I diabetes, according to the definitions adopted. The rate of remission to a NIR state was consistent with that observed in the open studies. Observations on the insulin-secretory status, and on the later clinical course of patients in remission in this study, were not presented.

In the Canadian-European Randomized Control Trial with ciclosporin, 188 patients with clinically diagnosed IDDM were entered within 14 weeks of the onset of classical symptoms, and with no more than 6 weeks of treatment with insulin [3]. Patients ranged in age from 9 to 35 years, but the great majority (86%) were 16 years or older. All had ketonuria at diagnosis. The study was designed to examine the effects of ciclosporin on remissions defined in terms of insulin treatment status, and on beta-cell function assessed in terms of plasma C peptide levels, through a minimum of 1 year. Two definitions of remission were employed. A clinical definition of the NIR remission required maintenance of target control of glycemia (self- and laboratory-determined values equivalent to plasma glucose concentrations <7.8 mmol/l before meals) without administration of insulin. A further compound definition of remission required either this NIR state, or the presence of glucagon-stimulated C peptide levels in the plasma >0.6 nmol/l, a value achieved by $>95\%$ of normal subjects. The patients and those managing the administration of insulin were blind to the treatment. The initial dose of ciclosporin was 10 mg/kg/day, and this was adjusted to maintain trough levels of the drug equivalent to 400–600 ng/ml of whole blood (polyclonal RIA). The dose was reduced if serum creatinine levels increased by more than 50% of the entry values, or in response to development of other unacceptable adverse effects. Placebo was substituted for ciclosporin in patients in the active treatment group who showed confirmed glucagon-stimulated C peptide levels <0.1 pmol/l. Glycemic control was monitored by serial determinations of glycated hemoglobin levels, and the insulin-secretory status was assessed in terms of glucagon-stimulated C peptide levels in the plasma at 3-month intervals.

Glycated hemoglobin levels did not differ in the two treatment groups through the period of study. A highly significant increase in the rate of remission to the NIR state was observed in the ciclosporin-treated group (fig. 1). During the first 3 months clinical remissions to the NIR state occurred at similar rates in the two treatment groups, but thereafter NIR remissions accumulated to 38% of ciclosporin-treated patients, compared with 19% of placebo-treated patients. However, by 12 months, the rates had declined to 24 and 9%, respectively. Multiple logistic regression analysis adjusting for differences in baseline variables showed that the differences between the remission rates by both definitions employed were highly significant. Further analysis of the results indicated a striking relationship between the duration of symptoms of diabetes at entry and the response to ciclosporin (table 1). Thus, the development of clini-

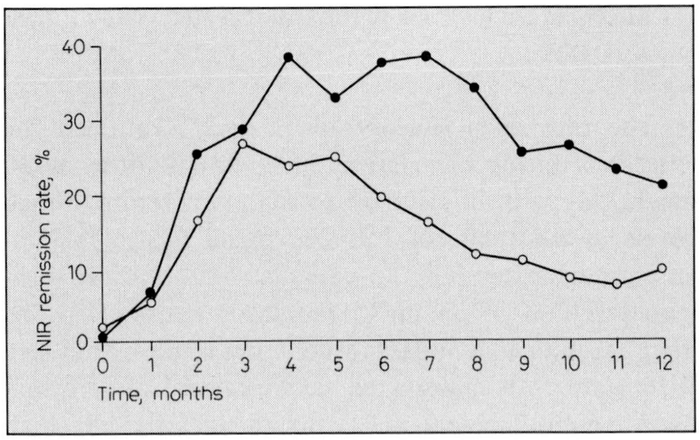

Fig. 1. NIR remission rates in the ciclosporin-treated (●) and placebo-treated (○) groups of patients at intervals through 1 year.

Table 1. Effect of duration of diabetes and ciclosporin on remissions

Outcome		6 months				12 months[2]			
		short[1]		long		short[1]		long	
		%	n	%	n	%	n	%	n
NIR or	CS	71.1	38	38.2	55	39.5	38	28.3	53
C peptide	PL	31.6	38	30.4	56	16.2	37	23.6	55
≥0.6 nmol/l	p	0.001		0.39		0.02		0.58	
	p diff.[3]		0.02				0.15		
NIR	CS	55.3	38	27.3	55	31.6	38	18.9	53
	PL	13.2	38	23.2	56	2.7	37	14.5	55
	p	0.001		0.62		0.001		0.55	
	p diff.[3]		0.003				0.06		

CS = Ciclosporin; PL = placebo.

1 Short = ≤6 weeks symptoms and ≤2 weeks insulin, to entry.

2 The minor discrepancies between numbers of subjects in the two response groups are accounted for by data falling outside predetermined time windows.

3 p value for difference in treatment effect between short-duration and long-duration subsets.

cal remissions in the ciclosporin-treated group was largely confined to the 40% of patients who entered the study with less than 6 weeks of symptoms and with less than 2 weeks of administration of insulin. The rate of remission to the NIR state at 1 year was 32% in ciclosporin-treated patients with 'short-duration' diabetes as defined, compared with 3% in placebo patients with longer-duration diabetes. The characteristics of patients with short- and longer-duration diabetes in the two treatment groups, including the levels of glucagon-stimulated C peptide in the plasma, were not otherwise different. With respect to the effect on beta-cell function, there was a significantly greater increase in the response of plasma C peptide to stimulation with glucagon by 90 days in the ciclosporin-treated patients (fig. 2). The enhanced beta-cell response continued at approximately the same level through the year in the ciclosporin-treated patients; in the placebo-treated group, however, the enhancement disappeared within 180 days, and the responses then declined below entry values.

This trial confirmed the efficacy of ciclosporin in induction of clinical remissions in IDDM of recent onset, and demonstrated for the first time that this is associated with enhancement and preservation of endogenous secretion of insulin. The anticipated clinical and biochemical adverse effects of ciclosporin were consistent with those in the open studies, which employed similar doses of ciclosporin. The changes in calculated creatinine clearance agreed well with direct measurements of glomerular filtration rate in subsets of patients; in the ciclosporin-treated group, the mean reduction of glomerular filtration rate in terms of calculated creatinine clearance was 20%, and by direct measurement was 21%.

Toxicity of Ciclosporin

The adverse effects of ciclosporin have been assessed in long-term experience in organ transplantation, and in shorter-term experience of experimental therapies in autoimmune diseases. The data indicate a very low risk of lymphoma or carcinoma, and a very high incidence of functional renal impairment. Thus, the nephrotoxicity of ciclosporin remains a major concern with respect to its use in IDDM. The immunosuppression programs first employed in studies in diabetes were similar to long-term regimens with ciclosporin in kidney transplantation. In that experience through many years there appeared no clear evidence of progressive or irreversible nephrotoxicity with ciclosporin [8]. However, more recently it has been reported that histological damage attributable to ciclosporin in the transplanted kidney is related to the duration of administration of the drug [9].

In view of the importance of this question, kidney biopsies from 17 patients taking part in the Canadian European trial were evaluated together with 23 biopsies from other patients in studies using similar protocols in France and Canada. These showed changes associated with administration of ciclosporin, though not specifically attributable to the drug [3]. The conditions quantified were glomerular obsolescence or segmental focal sclerosis, arteriolopathy, tubular atrophy and interstitial fibrosis, which were rated on 5-point scales. The median score for glomerular change was zero, but 1 biopsy showed abnormalities in more than 10% of glomeruli. The commonest finding was interstitial fibrosis, with a median score of 'slight', but the score was 'moderate' in 8 specimens. The median score for tubular atrophy was 'minimal-slight', but this lesion was rated moderate in 7 specimens. This last abnormality was the only one that correlated with trough levels of ciclosporin in the blood. The incidence of tubular atrophy showed a minimal median score in biopsies from patients in whom the mean ciclosporin trough levels had been less than the approximate median value for the studies. The findings suggested that an upper limit of 400 ng/ml whole blood for the 12-hour trough levels of ciclosporin should minimize the occurrence of structural damage. This expectation is supported by results of the recent open study with ciclosporin in children in France discussed above [6], in which the upper limit of ciclosporin level in the blood was 350 ng/ml, and in which there was no significant functional effect on the kidney, with no definite histological abnormality in the 19 kidney biopsies obtained, while the NIR remission rate was comparable to that observed with higher doses of ciclosporin. However, it should be noted that confirmation of induction of remissions with the lower dose of ciclosporin in a randomized control trial is awaited, and that the effect of lower doses of ciclosporin on beta-cell function awaits documentation. Nevertheless, it appears that doses of ciclosporin which are not demonstrably nephrotoxic through the relatively short periods of study employed to date can induce NIR remissions. Studies with the higher doses of ciclosporin formerly used have demonstrated complete reversibility of the functional biochemical and hematological adverse effects of the drug [10].

2
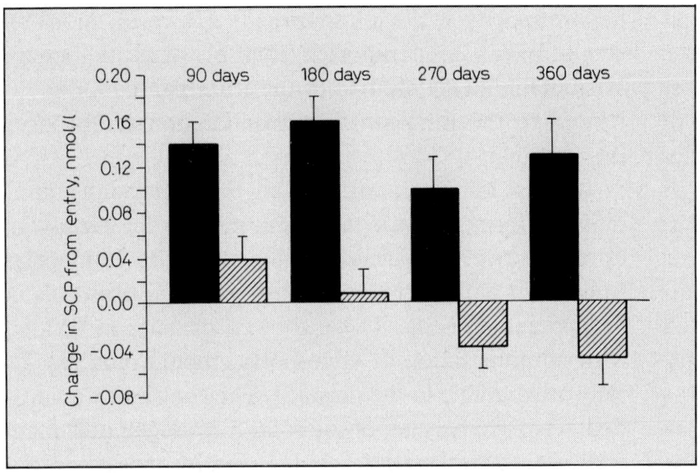

3
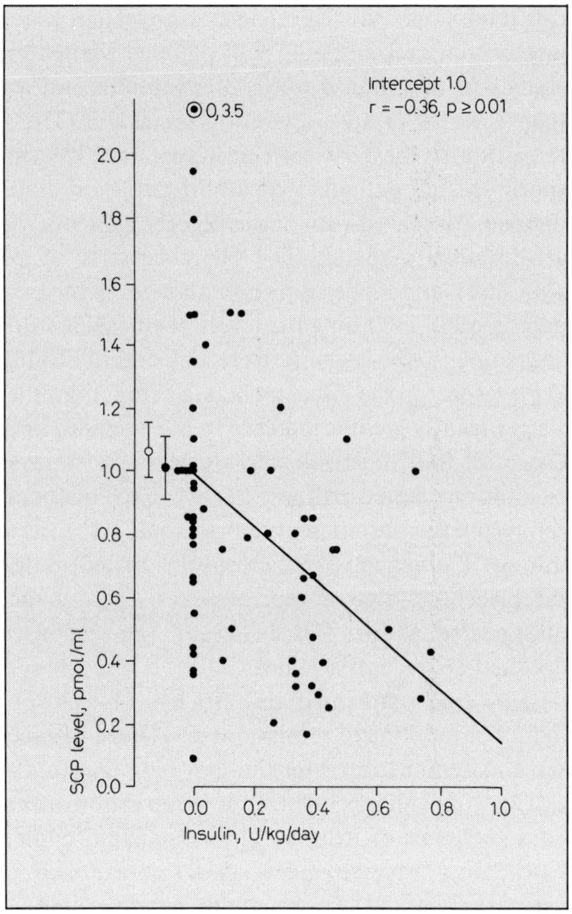

Fig. 2. Mean changes (± SE) from baseline values in stimulated insulin-connecting peptide (SCP) levels in plasma in ciclosporin-treated (■) and placebo-treated (▨) patients. Differences between changes in the two groups were statistically significant at all times ($p < 0.01$).

Fig. 3. Stimulated plasma C peptide (SCP) levels and concurrent daily insulin dose in individual patients after 3 months of treatment with ciclosporin. The solid symbol with bars indicates the mean ± SEM for all NIR patients (1.01 ± 0.10 pmol/ml); the open symbol with bars indicates the mean ± SEM in 22 normal subjects (1.07 ± 0.09 pmol/ml). From Canadian Open Studies.

IDDM in Remission: the Endocrine-Metabolic Status and Its Possible Implications

Studies of the insulin-secretory capacity based on determination of responses of C peptide to various stimuli, and more limited information regarding blood levels of immunoreactive insulin itself, show that patients in NIR remission as defined can maintain normal-range glycated hemoglobin levels, and that many also exhibit glucagon-stimulated C peptide levels in the blood within the normal range. In NIR patients in the Canadian-European randomized trial, the mean values for glucagon-stimulated C peptide were approximately 50% of the mean normal level in both ciclosporin- and placebo-treated groups [3]. Moreover, determinations of glucagon-stimulated C peptide in all patients receiving ciclosporin in the Canadian open study during the 1st year of observation showed that those in NIR remission exhibited a mean glucagon-stimulated plasma C peptide response that was not distinguishable from normal [9] (fig. 3). However, it is also clear from these data that there was considerable overlap of glucagon-stimulated C peptide levels between patients who entered NIR remissions and those who were unable to discontinue insulin. These findings suggest that changes in insulin sensitivity may play a role not only in determining which patients enter NIR remission, but may also affect the longer-term stability of remissions. Preliminary studies of insulin sensitivity in patients with NIR remissions during administration of ciclosporin support the suggestion that declining insulin sensitivity is important in relapse to renewed requirement for insulin [11].

Endocrine Responses to Nutrients in Remission-Phase Diabetes mellitus

There have been few studies of the physiological responses of the endocrine pancreas in patients under conventional treatment for IDDM during the remission phase. Preliminary reports of studies in patients with ciclosporin-induced remissions have appeared [12]. It was shown that the release of C peptide and free immunoreactive insulin into the blood is virtually absent in response to an intravenous infusion of glucose alone, in patients in NIR remission with normal levels of glycated hemoglobin in the blood. On ingestion of glucose, substantial increases of insulin and C peptide in the blood occurred. By contrast, the responses of insulin and C peptide following ingestion of a standardized mixed meal (Sustacal) were greater and were associated with only a minor degree of glucose intolerance. These patients showed obvious intolerance with intravenous or oral glucose, but glucose tolerance after ingestion of the mixed meal was only mildly impaired, and was consistent with the observed maintenance of normal-range glycated hemoglobin levels under physiological conditions.

Thus, it appears that patients in NIR remission exhibit the same abnormality of response to glucose as that described in the preclinical phase of the disease in high-risk siblings of diabetic patients. This suggests either that this abnormality of beta-cell function persists, and that patients in remission revert to a condition similar to that of patients in the preclinical phase [13], or that the defective insulin response to glucose in remission-phase patients is attributable wholly or in part to other effects of ciclosporin, as discussed below.

Endocrine Metabolic Effects of Ciclosporin

At present it seems unlikely that direct endocrine-metabolic effects of ciclosporin are important in determining the metabolic status of ciclosporin-treated patients with IDDM receiving this agent. Thus, intravenous glucose tolerance and the associated insulin response in patients with multiple sclerosis receiving doses of ciclosporin similar to those employed in studies in diabetes are normal [14]. Studies of insulin receptor function with erythrocytes have demonstrated no major abnormality in ciclosporin-treated patients with IDDM [15]. Moreover, the preliminary observations on insulin sensitivity in ciclosporin-treated patients with IDDM cited above indicate that this function is normal in patients in NIR remission [11]. Thus, it appears that ciclosporin-induced remissions of IDDM are characterized by a continuing abnormality of the beta-cell response to stimulation with glucose, with well-preserved responses to mixed meals, presumably mediated by non-glucose nutrients and/or by enteroinsular mechanisms. These patients can maintain normal-range glycated hemoglobin levels, under physiological conditions. Thus they differ from 'C-peptide-negative' insulin-dependent diabetics achieving similar control as a result of intensive insulin therapy, who manifest elevated fasting and integrated diurnal plasma levels of free immunoreactive insulin [16]. Clearly, much remains to be learned about the physiology of 'spontaneous' and treatment-induced remission of IDDM, with respect to the characteristics that distinguish NIR subjects from those who continue to require insulin, both in regard to the secretion patterns and efficacy of the hormone.

Clinical Significance of Preservation of Beta-Cell Function in IDDM

It is important to consider the possibility that long-term clinical benefit may result from preservation of levels of beta-cell function that are not sufficient to allow withdrawal of insulin therapy. While the NIR remission has generally been the goal in studies of experimental immunotherapy in IDDM in man, it is apparent that preservation of the characteristics of remission-phase IDDM might offer clinical benefit even with continued administration of insulin, provided the therapy inducing this effect were safe. In remission-phase IDDM, maintenance of virtual normoglycemia with low-dose insulin treatment is possible in many patients. It has been pointed out that the levels of insulin in the blood in patients in NIR remission are in the normal range, and unpublished data of the authors indicate that this is also true of patients undergoing the low-dose insulin therapy required for strict glycemic control under these conditions. Moreover, it seems that endogenous insulin secretion contributes to the efficacy of insulin in controlling glycemia, even when residual beta-cell function is low [17]. Is also appears that modest preservation of beta-cell function is associated with reduced doses of insulin and relatively low levels of glycated hemoglobin in remission-phase IDDM [18]. Furthermore, there is evidence that patients receiving low doses of insulin are at low risk for the long-term complications of IDDM [19], and that demonstrable residual beta-cell function is present in many patients in this category decades after diagnosis [20]. It therefore appears that preservation of definable levels of C peptide in the blood may be associated with reduced morbidity and mortality in insulin-treated IDDM. This possibility indicates the need for a continuing search for immunomodulatory interventions that protect residual beta-cell function after the onset of overt IDDM, when the pathophysiology in the treatment-enhanced remission appears to be quite similar to that of preclinical type I diabetes at the stage of loss of the beta-cell response to direct stimulation by glucose.

References

1 Stiller CR, Dupre J, Gent M, et al: Effects of cyclosporine immunosuppression in insulin-dependent diabetes mellitus of recent onset. Science 1984;223:1362–1367.

2 Sichett E, Daneman D, Clarson C, et al: Factors affecting and patterns of residual insulin secretion during the first year of type I (insulin-dependent) diabetes mellitus in children. Diabetologia 1987;30:453–459.

3 Canadian-European Diabetes Study Group: Cyclosporin-induced remission of IDDM after early intervention: Association of 1 year of cyclosporin treatment with enhanced insulin secretion. Diabetes 1988;37:1574–1582.

4 Dupre J, Stiller CR, Gent M, et al: Effects of immunosuppression with cyclosporine in insulin-dependent diabetes mellitus of recent onset: The Canadian Open Study at 4 months. Transplant Proc 1988;20(suppl 4):184–192.

5 Stiller CR, Dupre J, Gent M, et al: Effects of cyclosporine in recent onset juvenile type I diabetes. Impact of age and duration of disease. J Pediatr 1988;111:1069–1072.

6 Bougneres PF, Carel JC, Castano L, et al: Factors determining very early remission of type I diabetes in children treated with cyclosporin A. N Engl J Med 1988;318:663–671.

7 Feutren G, Papoz L, Assan R, et al: Cyclosporin increases the rate and length of remissions in insulin-dependent diabetes of recent onset. Lancet 1986;ii:119–129.

8 The Canadian Multicentre Transplant Study Group: A randomized clinical trial of cyclosporine in cadaveric renal transplantation. N Engl J Med 1986;314:1219–1225.

9 Ruiz P, Kolbeck PC, Scroggs MW, et al: Association between cyclosporine therapy and interstitial fibrosis in renal allograft biopsies. Transplantation 1988;45:91–95.

10 Dupre J, Stiller CR, Gent M, et al: Clinical trials of cyclosporin in IDDM. Diabetes Care 1988;11(suppl 1):37–44.

11 Hramiak I, Finegood D, Dupre J: Insulin sensitivity in cyclosporine treated type I diabetes in remission (abstract 63). 48th Annu Meet ADA, New Orleans, June 1988. Diabetes 1988;37(suppl 1): 16A.

12 Dupre J, Stiller CR, Jenner M, et al: Responses to nutrients in non-insulin requiring (NIR) remission of type I diabetes during administration of cyclosporin (abstract): 47th Annu Meet ADA, Indianapolis, June 1987. Diabetes 1987;36(suppl 1):74A.

13 Srikanta S, Ganda OP, Jackson PA, et al: Type I diabetes mellitus in monozygotic twins: Chronic progressive beta cell dysfunction. Ann Intern Med 1983;99:320–326.

14 Robertson RP, Franklin G, Nelson L: Intravenous glucose tolerance and pancreatic islet B cell function in patients with multiple sclerosis during two years' treatment with cyclosporine. Diabetes 1989;38:58–64.

15 Behme MT, Dupre J, Stiller C: Effect of cyclosporine on insulin binding to erythrocytes in type I diabetes mellitus of recent onset. Clin Invest Med 1988;11:113–122.

16 Bergenstal RM, Dupre J, Lawson PM, et al, for the KROC Collaborative Study Group: Observations on C-peptide and free insulin in the blood during continuous subcutaneous insulin infusion and conventional insulin therapy. Diabetes 1985; 34(suppl 3):31–36.

17 Rodger NW, Dupre J, Canny CLB, et al: Continuous subcutaneous insulin infusion in adults: Glycemic advantage is predicted by venous plasma C-peptide concentrations. Diabetes Care 1985;8:447–455.

18 The DCCT Research Group: Effects of age, duration, and treatment of IDDM on residual B-cell function: Observations during eligibility testing for the DCCT. J Clin Endocrinol Metab 1987; 65:30–36.

19 Deckert T: Influence of supervision and endogenous secretion of insulin on the course of IDDM. Acta Endocrinol 1980;94(suppl 238):31–38.

20 Eff CL, Faber O, Deckert T: Persistent insulin secretion assessed by plasma C-peptide estimation in long-term juvenile diabetes with a low insulin requirement. Diabetologia 1978;15:169–172.

Dr. J. Dupre
University Hospital
PO Box 5339
London, Ont. N6A 5A5 (Canada)

Discussion

H.B. Mortensen, Denmark: Did you investigate kidney during treatment with ciclosporin?

J. Dupre, Canada: I did not mention that, not to be evasive but because it would take a long time and these data have already been published. There was the expected approximate 30% decline in glomerular filtration rate which was verified in more than half the patients by direct measurements. Our experience being with the doses cited (with 1 exception out of 200 patients) is that a complete reversal to the expected predicted normal corrected rate of glomerular filtration rate was found on discontinuation. From renal biopsy data from patients on this dosage, an incidence of about 15% of structural changes, that are almost certainly due to the ciclosporin, has been found and that has been the reason for the changes in the dosage we and our French colleagues have adopted.

Horm Res 1990;33:159

General Discussion

A.L. Drash, USA: I said that Dr. Dupre and his group brought us into the ciclosporin age. I think we are in the last stages of the ciclosporin era in terms of its application to diabetes. There is a new very potent immunosuppressive in the field that you need to be informed about, which is called by code FK506. It is a Japanese product (Fujisawa, Osaka), an immunosuppressive agent derived from the fungus of streptomycetes. It is interesting that the Japanese equivalent of the American Food and Drug Administration have refused to allow its use in humans in Japan. However, it is being used in Pittsburgh very actively by Tom Starzl, who is a transplant surgeon. The first scientific report appeared in the New York Times about 2 weeks ago (mid-October 1989) in which the scientific editor reported on 120 cases in which this compound has been used in liver transplantations. The first true scientific report is in the current issue of Lancet [Starzl et al., Lancet 1989;ii:1000–1006], entitled 'FK506 for Liver, Kidney and Pancreas Transplantations'. Dr. Starzl feels that this is a miracle drug. His hyperbole is that it is 500 times as effective as ciclosporin and has no adverse side effects. We have a great deal of hesitancy in accepting that but we are under considerable pressure to move into clinical trials in newly diagnosed diabetics as Dr. Starzl feels that this compound will have great applications across the world, not only in transplantations but in autoimmune diseases in general. Certainly from the Pittsburgh pediatric perspective, it will not be given to diabetic children until we are fully convinced both of its effectiveness and safety and we are currently doing studies in the BB rat with this compound.

P. Czernichow, France: It is difficult to speak about results that have not yet been completely analyzed and Dr. Bougneres (the leader of the study) would be the best person to answer but as he is not here. I shall give you some information on the preliminary results we have. In June 1988, Dr. Bougneres (Hôpital St-Vincent-de Paul) and Dr. Lévy-Marchal (Hôpital Debré) started a double-blind randomized study involving 60 children. Thirty were given placebo and 30 ciclosporin. The criteria of inclusion were such that we hoped that beta-cell destruction was moderate, like history of polyuria, moderate loss of weight and blood pH above 7.25. The children received 6 mg/kg of ciclosporin aiming at a blood level of below 50 mg/ml of ciclosporin and with other safety taps which I will not describe. Criteria for total remission were quite severe: no insulin at all, HbA_1 below 7%, fasting blood glucose of 140 mg/dl and postprandial blood glucose of <200 mg/dl. Up to 4 months, the results were identical to what you have seen in the adult studies: 40% of patients on ciclosporin and 20% in the placebo group were in total remission. After a year of treatment surprisingly 13 patients in the placebo group and 17 patients on ciclosporin were in total remission. We have learnt from this study the importance of a control group; I also seriously doubt that with the use of a low dose of ciclosporin (6 mg/kg) as we have been using, we are going to get significant results for cure of insulin-dependent diabetes mellitus.

J. Dupre, Canada: Do you know about the C-peptide response in these trials? Is there the same dissociation as in the clinical response?

P. Czernichow, France: I do not know.

J. Nerup, Denmark: I think it was clear from Dr. Elias' presentation that T killer cells are very important in the process leading to diabetes and this is very pertinent knowledge for the last remarks by Dr. Dupre, since the major effect of ciclosporin is to block the transcription of a series of genes in these lymphocytes.

Epilogue

Horm Res 1990;33:160

0301-0163/90/0334-0160$2.75/0

Memories of the Third Joint Meeting

G.P. August

Department of Endocrinology and Metabolism, Children's Hospital, Washington, D.C., USA

Antibodies, we hear, amok have run;
Beta-cells, their targets, simply succumb.

Control techniques, we still greatly debate;
Diabetes' scourges, though, can abate.

Endocrine, autocrine, paracrine all,
Factors yet to be discovered call.

Growth was the focus of much attention;
Height ultimately is our intention.

Intron, exon, to learn of these is fun?
Jibberish for some; but an RO-1.

Mediators, are they unitary?
Nuclei, target itinerary.

Oscillating neurons, what have they wrought?
Pulsatile secretions, now are brought.

Quirks of behavior, to what are ascribed?
Rhythms, both long and short, we heard described.

Somatomedins doth commence the feat;
Testosterone plus estrogen complete.

Under Israeli skies of white and blue
Vistas unfold; ancient dreams renew.

Wonders of entertainment we behold;
Xerxes, himself, could never be so bold.

Yerushalayim's feast of knowledge clear;
Zion, todah; shalom until next year.